MICROBIOLOGY IN AGRICULTURE AND HUMAN HEALTH

Edited by **Mohammad Manjur Shah**

Microbiology in Agriculture and Human Health
http://dx.doi.org/10.5772/59645
Edited by Mohammad Manjur Shah and co-editors Tanveer Beg, Md. Mahamood,
L. Jayentakumar Singh, L. Bina Chanu

Contributors

Wanjun Zhang, Hiroo Uchiyama, Huhe -, Yuanbai Pan, Galina Novik, Victoria Savich, Elena Kiseleva, Hongxuan He, Wen Su, Debora Ferreira Barreto-Vieira, Ortrud Monika Barth, Nora-Hilda Rosas-Murrieta, Sandra R. Reyes-Carmona, Rebeca D. Martínez-Contreras, Nancy Martínez-Montiel, Lourdes Millán-Pérez-Peña, Irma Herrera-Camacho, Guadalupe Rojas-Sánchez, Mohammad Manjur Shah, Loukrakpam Bina, N. Mohilal

CBS, Edition **2017**

Published by InTech

Janeza Trdine 9, 51000 Rijeka, Croatia

Publishing Process Manager Sandra Bakic

Technical Editor InTech DTP team

Cover InTech Design team

Additional hard copies can be obtained from orders@intechopen.com

Microbiology in Agriculture and Human Health, Edited by Mohammad Manjur Shah

p. cm.

ISBN 978-953-51-2144-2

INTECH

open science | open minds

Contents

Preface

Microbiology involves the study of microscopic living organisms that consist of bacteria, fungi, algae, protozoa, and infectious agents such as viruses. The study also involves their form, structure, reproduction, physiology, metabolism and classification as well. Most of them are unicellular and all the life processes are performed by a single cell. They are associated with the health and welfare of human beings; and while some of them are beneficial, others cause diseases.

Among the biological sciences, microbiology has established itself a place in the current century. Microorganisms also provide experimental models in various research activities including molecular biology and biotechnology, and in turn, provide answers to numerous fundamental questions in genetics / metabolism, and cell form and function.

Genomic tools are opening innovative and therapeutic advancements in the study of the genome of microorganisms. Environmental, food, and medical microbiology research is evolving to address global issues created by surging populations challenged by climate and water resources. This book addresses a wide array of subjects in novel microbiological research and study, including discussion of environmental, medical, pharmaceutical, veterinary, and agricultural research. It includes discussion on soil, water and bio-deterioration. Discussion ranges from the basics of microbial genetics, growth, metabolism and physiology.

Infectious disease research has been emphasized within the context of medical microbiology, cellular microbiology and bioprocess engineering. This book also addresses industrial microbiology applications and techniques, and the role of microbial ecology in global planning.

The book audience and contributors consists of microbiologists including bacteriologists, virologists, mycologists, parasitologists and molecular microbiologists, and advanced students of those subjects.

The book is presented in six chapters comprising of two sections. The first section deals with Microbiology and Agriculture, and the second section deals with Microbiology and Human Health.

While editing this book, each chapter has been the sole responsibility of each contributing author and their respective co-authors. All editors read and critiqued all the chapters and finally accepted after obtaining compliance on our suggested comments / improvements. The book is expected to attract wide audience from various fields of biological sciences in general, and microbiologists in particular.

We would like to thank the Publishing Process Manager of the book for overall cooperation and keeping full confidence to me while discharging my responsibility.

Editor:
Dr. Mohammad Manjur Shah
Department of Life Sciences,
Section of Parasitology,
Manipur University, India

Co-editors:
Tanveer Beg,
Jazan University, Saudi Arabia
Md. Mahamood,
Aligarh Muslim University, India
L. Jayentakumar Singh,
Manipur University, India
L. Bina Chanu,
Manipur University, India

Section 1

Microbiology and Agriculture

Chapter 1

Evaluation of the Efficiency of Some Antagonistic *Trichoderma* spp. in the Management of Plant Parasitic Nematodes

L. Bina Chanu, N. Mohilal and M. Manjur Shah

Additional information is available at the end of the chapter

http://dx.doi.org/10.5772/60082

Abstract

Plant parasitic nematodes cause great economic losses to agricultural crops worldwide. They along, with their hosts, are not isolated in the ecological system, but are strongly influenced by antagonists, parasites and pathogens. Though pesticides appear to be the most economical and efficacious means of controlling plant pathogens, toxicological, environmental and sociological concerns have led to drastic reductions in the availability of efficient commercial nematicides. These restrictions have forced farmers to look for an integral system that makes use of other means of disease control. Species of spiral nematodes, *Helicotylenchus* and *Scutellonema*, were among the most abundant plant parasitic nematodes of the mulberry plant. Eco-friendly control of the parasitic nematodes could be achieved by means of endoparasitic fungi (like *Hirsutella, Meria, Nematophthora* and *Nematoctonus*), trapping fungi (like *Arthrobotrys* and *Duddingtonia*) or parasitic fungi (like *Paeceilomyces lilacinus*). During the course of this present work, Trichoderma Pers. Ex. Fr. was found to be one of the most effective fungi in controlling the eggs and J_2 of *Meloidogyne javanica*. The present study outlines the comparative efficacy of five *Trichoderma* species (*T. viride, T. harzianum, T. longibrachiatum, T. koningii and T. hamatum*) against *Helicotylenchus* sp. and *Scutellonema* sp. The study also outlines the effect of *Trichoderma viride* Persoon on *Scutellonema* spp. and *Helicotylenchus* sp., effect of *Trichoderma harzianum* Raifae on *Scutellonema* sp. and *Helicotylenchus* sp., effect of *Trichoderma longibrachiatum* Rifai on *Scutellonema* sp. and *Helicotylenchus* sp., effect of *Trichoderma koningii* Oudeom on *Scutellonema* sp. and *Helicotylenchus* sp., and lastly effect of *Trichoderma hamatum* (Bonord) Bainier on *Scutellonema* sp. and *Helicotylenchus* sp.

Keywords: Plant parasitic nematodes, mulberry plant, fungus, antagonistic *Trichoderma* spp, biocontrol

1. Introduction

Plant parasitic nematodes cause great economic losses to agricultural crops worldwide. They along, with their hosts, are not isolated in the ecological system, but are strongly influenced by antagonists, parasites and pathogens. Though pesticides appear to be the most economical and efficacious means of controlling plant pathogens, toxicological, environmental and sociological concerns have led to drastic reductions in the availability of efficient commercial nematicides. These restrictions have forced farmers to look for an integral system that makes use of other means of disease control. This imperative approach involves a mixture of agro-technical, biological, chemical and genetic (breeding) means of plant disease control [20, 24, 36]. Species of spiral nematodes, *Helicotylenchus* and *Scutellonema*, were among the most abundant plant parasitic nematodes of the mulberry plant. Reductions in length and weight of shoot, number and weight of leaves, and number of leaf buds were the characteristic symptoms of the infection of spiral nematodes [10]. Rao and Swarup [26] found stunting of the plants and reduction in fresh and dry weights of both shoot and root system in sugarcane due to *Helicotylenchus dihystera*. Besides chemicals, various researchers suggested other control measures in view of the need to replace highly toxic and potentially polluting chemicals used to control plant parasitic nematodes and fungi with less dangerous chemicals, or preferably with biological control agents and botanicals [21]). The discovery of new biocontrol agents and the demonstration of their value in reducing disease incidence and severity has opened promising new avenues for practical applications in agriculture as well as for promoting environmental safety [8]. Considerable efforts have been made by many researchers for the management of different plant parasitic nematodes with the use of *Trichoderma harzianum* [1 - 5, 23, 28, 33].

Eco-friendly control of the parasitic nematodes could be achieved by means of endoparasitic fungi (like *Hirsutella, Meria, Nematophthora* and *Nematoctonus*), trapping fungi (like *Arthrobotrys* and *Duddingtonia*) or parasitic fungi (like *Paeceilomyces lilacinus*). But there are problems in the culture of the fungi, such as unavailability of their host, and the generalist feeding nature of fungi that means they can become trapped on and digest beneficial as well as pest species of nematode. The general approach has been to go to locations where nematodes have reached high densities, and extract parasitized individuals from the soil. Then, the fungi were cultured and tested as parasites of the nematode pest. The mycoparasitic ability of *Trichoderma* sp. against soil-borne plant pathogens allows for the development of biocontrol strategies [11, 13, 14, 16, 24]). Windham *et al.* [40] reported reduced egg production in the root-knot nematode *Meloidogyne arenaria* following soil treatment with *Trichoderma harzianum* and *T. koningii* preparations. Combining *T. harzianum* with neem cakes reduced the population of citrus nematode, *Tylenchulus semipenetrans* [25]. Reduction of *M. javanica* infection with several isolates of *Trichoderma lingnorum* and *T. harzianum* has been reported [32]. *Trichoderma* may also promote plant growth [19].

During the course of this present work, *Trichoderma* Pers. Ex. Fr. was found to be one of the most effective fungi in controlling the eggs and J_2 of *Meloidogyne javanica*. The fungi is characterized by rapidly growing colonies bearing tufted or postulate, repeatedly branched conidiophores with lageniform phialides and hyaline or green conidia borne in slimy heads. They can be cultured and isolated from any type of soil. Considering the importance of the fungal genus containing species that have the potential for economic impact, the present study was carried out to determine the comparative efficacy of five *Trichoderma* species (*T. viride*, *T. harzianum*, *T. longibrachiatum*, *T. koningii* and *T. hamatum*) against *Helicotylenchus* sp. and *Scutellonema* sp.

2. Materials and Methods

2.1. Extraction of nematodes

Soil samples from around the rhizospheric regions of mulberry plants were collected and processed through Cobb's sieving and decanting method followed by Baerman's funnel technique [38]. The nematodes were observed under stereoscopic microscope and were counted using a Syracuse counting disc.

2.2. Isolation and enumeration of *Trichoderma* sp. from soil

The fungi were isolated through the serial soil dilution plate method [39]. Then, 10 g of oven dried fungi was added to a sterile Erlenmeyer flask with 90 ml sterile water, and the mixture was stirred with a magnetic stirrer for 20-30 minutes. A blender was used for blending the samples. While the suspension was in motion, 10 ml of solution was taken and added to 90 ml sterile water in a screw-cap flask or medicine bottle. It was shaken for one minute and 10 ml of the suspension was transferred to a 90 ml sterile water blank. The process was repeated until the desired dilution was obtained. Ten millilitres of soil solution was pipetted and mixed with 90 ml distilled water and marked to 10^{-3}. From 10^{-2} and 10^{-3} test tubes about 5 ml solution was added to culture media contained in four petri dishes (two of each) and kept at laminar flow for 3-4 days.

To facilitate uniform spreading of the suspension over Czapek Dox agar surface at pH 5.5, the plate was placed on a turntable and the suspension spread with a flamed L-shaped rod with one hand, while rotating the turntable with the other. To obtain distinct colonies, plates were prepared 2-3 days before use or placed for a few hours at 35 to 40° C after pouring to ensure a dry agar when the suspension was added. A water film on the freshly poured plates caused excessive spreading of organisms. The plate was incubated for a few days at 24-30° C and colony counted.

The composition of the culture media was as follows:

1. Sodium nitrate ($Na_2 NO_3$) - 1.0 g
2. Magnesium sulphate ($Mg SO_4 \cdot 7 H_2O$) - 0.5 g
3. Potassium chloride (K Cl) - 0.25 g
4. Potassium dihydrogen phosphate (KH_2PO_4) - 0.5 g

5. Ferrous sulphate ($FeSO_4.7\ H_{2O}$) - 0.5 g
6. Sucrose - 15.0 g
7. Agar agar - 10.0 g
8. Distilled water – 500 ml

A broth media was made by mixing together the substances excepting the agar. It was kept for 24 hours to dissolve the substances completely. Five to six drops of the broth were removed with a dropper into different autoclaved cavity blocks.

2.3. Inoculation of nematode and fungi

Five to six drops of the broth were removed with a dropper into different autoclaved cavity blocks. Then, 0.1 ml of selected *Trichoderma* spp. was transferred into the cavity block containing the broth. Next, 200 female *Scutellonema* spp. and *Helicotylenchus* spp. each were also inoculated into the cavity blocks. The cavity blocks containing the whole mixture were incubated at room temperature covered upside down by autoclaved petri dishes. Uninoculated nematodes on the broth were also kept as control. Observation of the nematodes under stereoscopic binocular microscope to record their mobility and fungal infection was done at each 12-hour interval. Each treatment was replicated three times.

3. Results and Discussion

The fungus attacked the nematodes though the production of conidia, sticky spores and mycelia, which on contact adhere to the cuticle and germinate, forming germ tubes that penetrated the nematodes. Then, they extended their hyphae inside the nematodes after penetration of the cuticle by conidia formation. These hyphae multiplied profusely. They inactivated the parasitic nematodes and immobilized them due to production of certain antibiotics and compounds. Observations of the immobility and parasitism of the nematodes due to the fungi were made every 12 hours. Each observation was replicated three times and the results are represented in tables 1–10. Photographs with graphs of parasitism are also provided in figures 1-32.

3.1. Effect of *Trichoderma viride* Persoon on *Scutellonema* spp. and *Helicotylenchus* sp. (table 1 and 2)

After 12 hours of inoculation, the fungus produced mycelium and conidia. The highest immobility was found at 108 hours of inoculation. The spores attached at the middle and anterior end of the body, and made the nematode immobile in the case of *Scutellonema* spp. After 24 hours of inoculation, many mycelium were found attached to the entire body of the nematode, due to which the body of the nematode was deformed and became shrunken, killing many of the nematodes. Body constrictions of nematodes might be due to the sucking of body contents by the fungus. Fifty percent of *Scutellonema* spp. out of 200 nematodes were immobilized by the fungus at 84 hours, and complete immobilization was observed at 108 hours of inoculation. In the case of *Helicotylenchus* spp., infection started within 24 hours of inoculation, and complete infestation occurred within 4-5 days of inoculation. The conidiophores of *Trichoderma viride* were less complicated; they formed aerial hyphae and coiled around the body of the nematode, producing smaller branches and ultimately forming a conifer-like branching system.

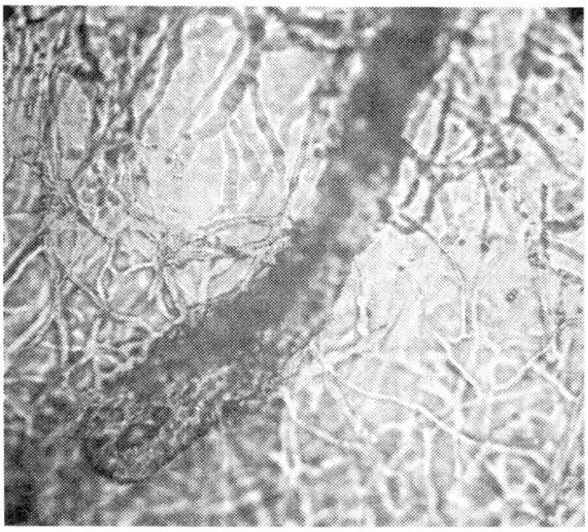

Figure 1. Effect of fungal inoculums of *Trichoderma viride* on *Scutellonema*

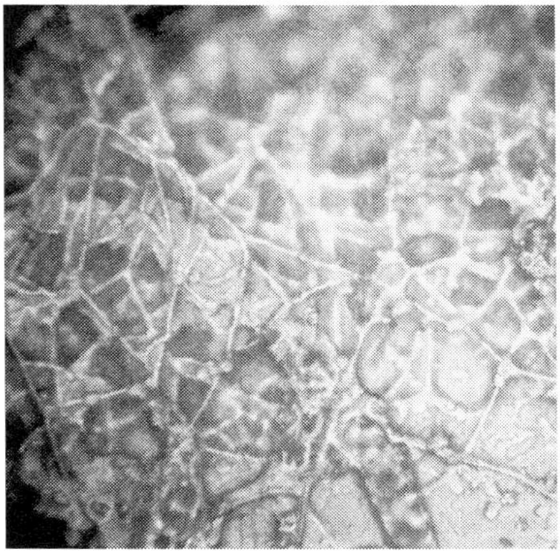

Figure 2. Effect of fungal inoculums of *T. harzianum* on *Helicotylenchus*

Observation	Active female nematodes	Non-active female nematodes	% of non-active female nematode	Control	
				Active	Non-active
12 hours	184	14.33	7.16	200	0
24 hours	174.33	25.66	12.83	200	0
36 hours	165.33	32.00	15.83	200	0
48 hours	148.66	52.66	26.33	200	0
60 hours	134.33	62.66	31.26	200	0
72 hours	120.66	84.00	42.00	200	0
84 hours	94.00	109.33	54.66	200	0
96 hours	48.66	154.0	77.0	200	0
108 hours	0.0	200.0	100	200	0
SE m±	19.33	19.67	9.84	0.0	0
C.D. at 0.05*	3.52	60.66	1.71	0.0	0

* Significant at 0.05 level of significance.

Table 1. Effect of fungal inoculums *Trichoderma viride* on the activities of *Scutellonema* spp.

Observation	Active nematode	Non-active female nematode	% of non- active nematode	Control	
				Active	Non-active
12 hours	200	0	0	200	0
24 hours	182	17	9	200	0
36 hours	172	27.66	13.83	200	0
48 hours	166	34	17	200	0
60 hours	145	54.66	27.33	200	0
72 hours	133	66.66	33.33	200	0
84 hours	126	74	37	200	0
96 hours	97	103	51.5	200	0
108 hours	46.33	153.66	76.83	200	0
120 hours	13	187	93.5	200	0
132 hours	0	200	100	200	0
SEm±	19.87	19.87	9.67	0.0	0
C.D. at 0.05*	2.760	2.760	1.38	0.0	0

* Significant at 0.05 level of significance.

Table 2. Effect of fungal inoculums *Trichoderma viride* on the activities of *Helicotylenchus* spp.

3.2. Effect of *Trichoderma harzianum Raifae* on *Scutellonema* sp. and *Helicotylenchus* sp. (table 3 and 4)

There was no infection after 24 hours of inoculation, but 3 % of the nematodes were immobile. Infection and parasitism of the nematode occurred after 48 hours of inoculation. The highest immobility was found at 108 hours of inoculation. Many mycelia grew over the body of the nematode. The conidiophores were seen to be multiple-branched, forming loose tufts which arose in distinct and continuous ring-like zones. The main branches, mostly in groups of two or three, stood at right angles, and the length increased with the distance from the tip of the main branch, giving a conical or pyramidal appearance. The body cuticle of the nematode was suppressed. The mycelia tip ran parallel to the nematode. There was rapid and excessive coiling on the target host. The mycelium coiled with its constricting networks of loops at the anterior region of the body and the head region, making constrictions that might be due to the sucking of body contents. After 96 hours of inoculation, there was complete immobilization of the nematodes. In the case of *Helicotylenchus*, the highest percentage of infection and immobility occurred during 96th hour of inoculation.

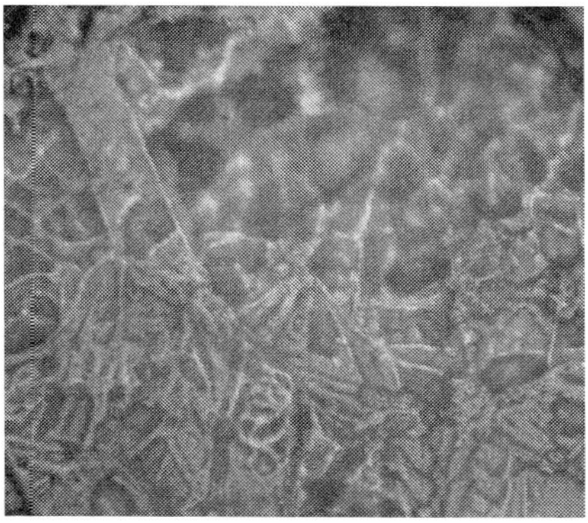

Figure 3. Effect of fungal inoculums of *T. harzianum* on *Helicotylenchus*

Observation	Active female nematodes	Non-active female nematodes	% of non-active female nematode	Control	
				Active	Non-active
12 hours	196.66	4.0	3.0	200	0
24 hours	198.66	5.0	3.5	200	0
36 hours	193.33	6.33	3.16	200	0
48 hours	161.33	36.0	17.66	200	0
60 hours	122.66	76.0	38.0	200	0
72 hours	104.0	90.0	45.0	200	0

Observation	Active female nematodes	Non-active female nematodes	% of non-active female nematode	Control	
				Active	Non-active
84 hours	89.66	104.66	52.33	200	0
96 hours	00	200	100	200	0
SEm±	22.88	22.47	11.15	0.0	0
C.D. at 0.05*	3.72	3.99	2.99	0.0	0

* Significant at 0.05 level of significance.

Table 3. Effect of fungal inoculums *Trichoderma harzianum* on the activities of *Scutellonema* spp.

Observation	Active nematode	Non-active nematode	% of non-active nematode	Control	
				Active	Non-active
12 hours	197	3	1.5	200	0
24 hours	193	7	3.5	200	0
36 hours	184.66	15.33	7.66	200	0
48 hours	162	38	19	200	0
60 hours	125.66	74.33	37.16	200	0
72 hours	108	92	47	200	0
84 hours	93	107	54	200	0
96 hours	53	147	73.5	200	0
108 hours	0	200	100	200	0
SEm±	21.34	19.32	10.67	0.0	0
C.D. at 0.05*	2.33	2.33	0.36	0.0	0

* Significant at 0.05 level of significance.

Table 4. Effect of fungal inoculums *Trichoderma harzianum* on the activities of *Helicotylenchus* spp.

3.3. Effect of *Trichoderma longibrachiatum Rifai* on *Scutellonema* sp. and *Helicotylenchus* sp. (table 5 and 6)

After 12 hours of inoculation, 4 % of the total nematode population was found to be immobile, with the highest immobility at 108 hours of inoculation. Infection started before 20 hours of inoculation. Hyphae of the fungus strain formed an appressorium-like structure in close contact with the nematode. They produced penetration holes in the cuticle of the nematode. The penetrated cuticle rapidly lost turgor and collapsed. At contact with the nematode cuticle, the hyphae branched dichotomously at the tip. The hyphae were not observed to coil around the nematode cuticle, and instead grew along the cuticle. However, penetration was not evident. Despite the absence of visible penetration, the nematode cuticle lost turgor pressure, wrinkled and collapsed. Finally, both the cuticle and body content of the nematode completely disintegrated. In the case of *Helicotylenchus*, the highest immobility was found at 60 hours of inoculation.

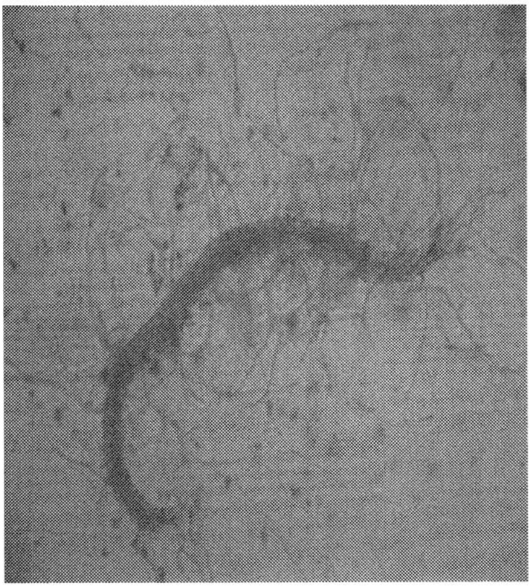

Figure 4. Effect of fungal inoculums of *T. longibranchiatum* on *Scutellonema*, whole body

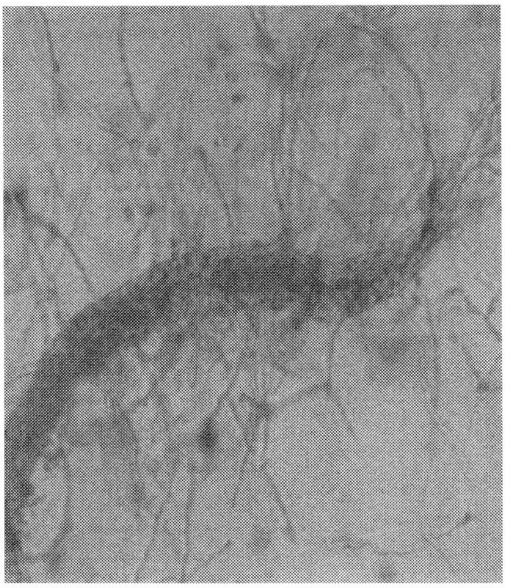

Figure 5. Effect of fungal inoculums of *T. longibranchiatum* on *Scutellonema*, head region

Observation	Active female nematodes	Non-active female nematodes	% of non-active female nematode	Control	
				Active	Non-active
12 hours	186.0	10.0	4	200	0
24 hours	180.0	18.0	9.0	200	0
36 hours	164	36	18.0	200	0
48 hours	154	46.33	22.33	200	0
60 hours	78.33	84.33	40.33	200	0
72 hours	83	121.33	60.13	200	0
84 hours	62	136.66	68.0	200	0
96 hours	27.33	173	86.53	200	0
108 hours	00	200	100	200	0
SEm±	21.70	21.81	10.97	0.0	0
C.D. at 0.05*	2.72	3.14	2.54	0.0	0

* Significant at 0.05 level of significance.

Table 5. Effect of fungal inoculums *Trichoderma longibrachiatum* on the activities of *Scutellonema* spp.

Observation	Active nematode	Non-active nematode	% of non-active nematode	Control	
				Active	Non-active
12 hours	193	7	3.5	200	0
24 hours	186	14	7	200	0
36 hours	174	26	13	200	0
48 hours	154	45	23	200	0
60 hours	83	117	58.5	200	0
72 hours	77	123	61.5	200	0
84 hours	52	147.66	83.5	200	0
96 hours	33	167	83.5	200	0
108 hours	12.66	187.33	93.66	200	0
120 hours	0	200	100	200	0
SEm±	22.24	22.05	11.02	0.0	0
C.D. at 0.05*	2.06	2.06	1.05	0.0	0

* Significant at 0.05 level of significance.

Table 6. Effect of fungal inoculums *Trichoderma longibrachiatum* on the activities of *Helicotylenchus* spp.

3.4. Effect of *Trichoderma koningii* Oudeom on *Scutellonema* sp. and *Helicotylenchus* sp. (table 7 and 8)

There was no effect during the first 12 hours of inoculation in the case of *Helicotylenchus* spp., but 7 % of *Scutellonema* spp. were immobilized during that time. After 24 hours of exposure, conidia attachment of the nematode was found. The conidia stuck towards the cephalic region and stylet of the nematode. Maximum immobility in the case of *Scutellonema* occurred at 144 hours of nematode exposure to the fungus, while it occurred at 168 hours of exposure in the case of *Helicotylenchus* sp. At 48 hours of exposure, hyphae formation was found around the body of *Helicotylenchus* and at the anterior and posterior part of the body of *Scutellonema* spp. The hyphae penetrated towards the body cuticle of the nematode and sucked the body contents, affecting the nematode. This might be attributed to the fungus's production of endo- and exochitinases by which hyphae penetration of the nematode cuticle was made possible.

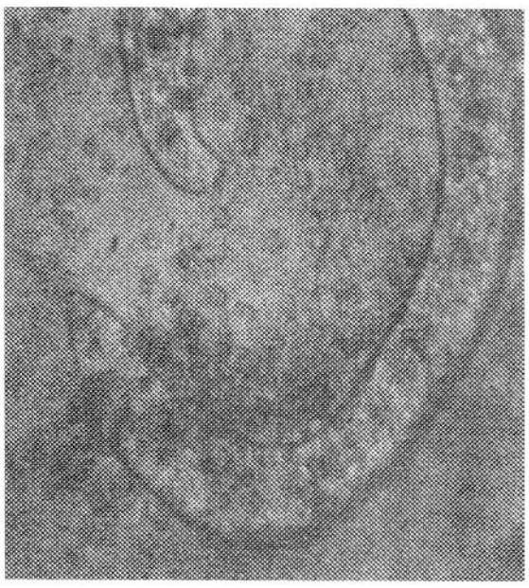

Figure 6. Effect of fungal inoculums of *T. koningii* on *Scutellonema*

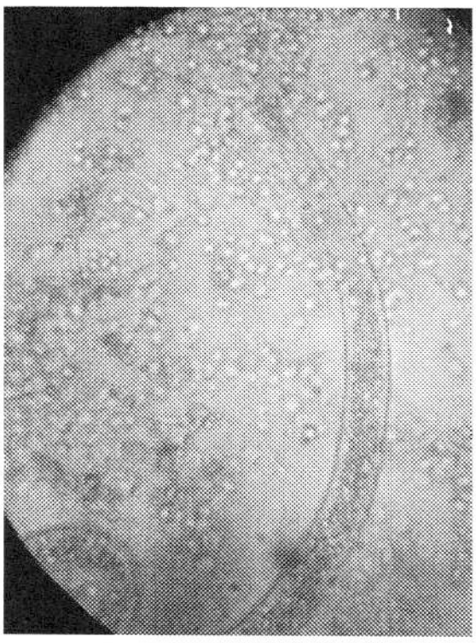

Figure 7. Effect of fungal inoculums of *T. koningii* on *Helicotylenchus*

Observation	Active female nematodes	Non-active female nematodes	% of non-active female nematode	Control	
				Active	Non-active
12 hours	186.0	13.0	7	200	0
24 hours	176.0	23	12	200	0
36 hours	166	33	17	200	0
48 hours	135	65	33	200	0
60 hours	127.0	73.0	37	200	0
72 hours	113.0	87.0	44.0	200	0
84 hours	97.0	100	50	200	0
96 hours	77	122	60	200	0
108 hours	65	132	65	200	0
120 hours	37	161	80	200	0
132 hours	13	185	92	200	0
144 hours	00	200	100	200	0
SEm±	17.269	17.192	8.483	0.0	0
C.D. at 0.05*	6.97	2.17	92.10	0.0	0

* Significant at 0.05 level of significance.

Table 7. Effect of fungal inoculum *Trichoderma koningii* on the activities of *Scutellonema* spp.

Observation	Active female nematodes	Non-active female nematodes	% of non-active female nematode	Control	
				Active	Non-active
12 hours	200	00	00	200	0
24 hours	198	1.33	0.66	200	0
36 hours	194	4.66	2.33	200	0
48 hours	179.66	16	7	200	0
60 hours	165	30	14	200	0
72 hours	133	62	30	200	0
84 hours	1235.33	70.0	34	200	0
96 hours	111	85	43.3	200	0
108 hours	95	101	51.3	200	0
120 hours	73.33	122	61.8	200	0
132 hours	60	135	68.3	200	0

Observation	Active female nematodes	Non-active female nematodes	% of non-active female nematode	Control	
				Active	Non-active
144 hours	44.66	150	75.93	200	0
156 hours	13	183.0	92	200	0
168 hours	00	200	100	200	0
SEm±	17.67	17.55	8.88	0.0	0
C.D. at 0.05*	26.088	3.089	1.891	0.0	0

* Significant at 0.05 level of significance.

Table 8. Effect of fungal inoculum *Trichoderma koningii* on the activities of *Helicotylenchus* spp.

3.5. Effect of *Trichoderma hamatum* (Bonord) Bainier on *Scutellonema* sp. and *Helicotylenchus* sp. (table 9 and 10)

There was no effect on the nematode *Scutellonema* spp. during the first 60 hours of exposure to the fungus, or 72 hours in the case of *Helicotylenchus* sp. Immobilizations of a few *Scutellonema* sp. were found at 72 hours of exposure, while this occurred at 96 hours of exposure in the case of *Helicotylenchus* sp. Hundred percent immobility of *Scutellonema* sp. was found at 300 hours of inoculation, and in the case of *Helicotylenchus*, it was found at 444 hours of exposure. Infection of *Scutellonema* sp. started after 68 hours of exposure, and 80 hours in case of *Helicotylenchus*. Direct growths of the mycoparasite from the body of the nematodes were observed. There was spore formation inside the body of the nematode and shrinkage of body contents occurred. *Trichoderma hamatum* produced aspersoria-like structures attached to the host cell wall. Subsequently several different types of interaction occurred. The fungus either grew parallel to and along the host hyphae or coiled around the host. In *Helicotylenchus* sp., the parasite penetrated into and grew within the cuticle. The cuticle became vacuolated, shrank, collapsed and finally disintegrated. The oesophageal part of the nematode had shrunken and the tail region was disintegrated into two, as in a fork.

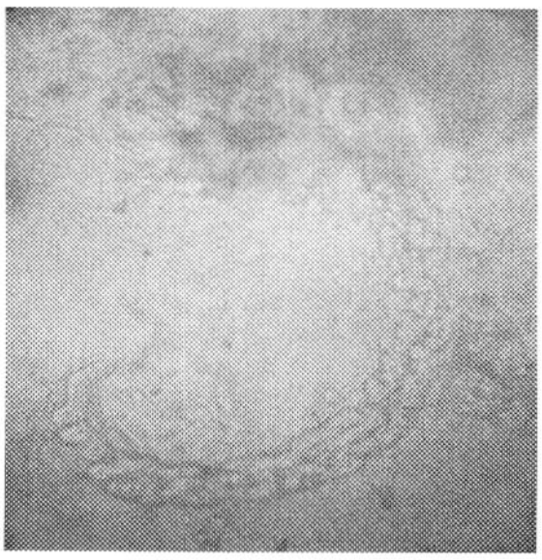

Figure 8. Effect of fungal inoculums of *T. hamatum* on *Scutellonema*

Figure 9. Effect of fungal inoculums of *T. hamatum* on *Scutellonema,* head region

Figure 10. Effect of fungal inoculums of *T. hamatum* on *Scutellonema,* tail region

Figure 11. Effect of fungal inoculums of *T. hamatum* on *Helicotylenchus*

Observation	Active female nematodes	Non-active female nematodes	% of non-active female nematode	Control	
				Active	Non-active
12 hours	200	00	00	200	0
24 hours	200	00	00	200	0
36 hours	200	00	00	200	0
48 hours	200	00	00	200	0
60 hours	200	00	00	200	0
72 hours	195	5	2.5	200	0
84 hours	184.33	15.66	7.8	200	0
96 hours	173.66	26.33	13.16	200	0
108 hours	165	35	17	200	0
120 hours	152	48	24	200	0
132 hours	146.33	53.66	26.83	200	0
144 hours	135.33	64.66	32.3	200	0
156 hours	123	77	38.5	200	0
168 hours	117	83	41.5	200	0
180 hours	105.33	94.66	44	198	2
192 hours	91	109	54.4	196	4
204 hours	80.33	123.66	61.83	194	6
216 hours	68.66	134.66	61.83	193	7
228 hours	54	146	73	191	9
252 hours	37	163	81.5	189	11
264 hours	26	174	87.33	187	13
276 hours	16.33	183.66	95.83	185	15
288 hours	8.33	191.66	95.83	183	17
300 hours	00	200	100	181	19
SEm±	4.907	13.98	7.14	1.25	1.25
C.D. at 0.05*	3.57	3.27	2.55	0.0	0.0

* Significant at 0.05 level of significance.

Table 9. Effect of fungal inoculum *Trichoderma hamatum* on the activities of *Scutellonema* spp.

Observation	Active female nematodes	Non-active female nematodes	% of non-active female nematode	Control	
				Active	Non-active
24 hours	200	00	00	200	0
48 hours	200	00	00	200	0
72 hours	200	00	00	200	0
96 hours	191.33	6.0	3	200	0
120 hours	180	15.66	7.66	200	0
144 hours	174.66	24	12	200	0
156 hours	150.66	50.66	25.33	200	0
180 hours	134.66	62.33	31.16	200	0
204 hours	122	77	48.5	194	6
228 hours	106	93	46.5	193	7
252 hours	85.33	119	59.16	191	9
276 hours	77	125	62.5	189	11
300 hours	63	140.66	70.33	187	13
324 hours	47	153	77.33	185	15
348 hours	30	171	85.63	183	17
372 hours	28.66	177	88.5	181	19
396 hours	17.66	186.33	94.83	179	21
420 hours	8.66	195.0	94.16	177	23
444 hours	00	200	100	176	24
SEm±	16.13	16.519	8.229	2.00	2.0
C.D. at 0.05*	2.86	3.61	7.08	0.0	0.0

* Significant at 5 % level of significance.

Table 10. Effect of *Trichoderma hamatum* on the activities of *Helicotylenchus* spp.

Several possible mechanisms have been suggested to be involved in *Trichoderma* antagonism, such as production of volatile or non-volatile antibiotics by the fungus [6], space- or nutrient- (carbon, nitrogen, iron, etc.) limiting factors that compete with the host [31], and direct mycoparasitism whereby the host cell wall is degraded by the lytic enzymes secreted by *Trichoderma* [9]. *Trichoderma harzianum* produced antibiotic 6-pentyl-α-pyrone, which had the dual effect of inhibiting pathogen growth and down-regulating genes for biosynthesis of trichothecenes, a class of mycotoxins with broad-spectrum antimicrobial activity [12]. *Trichoderma longibrachiatum* produced three main hydrolytic enzymes: protease, β-1, 3-glucanase and chitinase, which were involved in fungal cell wall degradation. *Trichoderma koningii* has also been found to produce cell wall degrading enzymes – chitinases, β-1, 3-glucanase and cellulose – which aid in the colonization of their host cells, while isonitrin, homothalin A, melanoxadin, trichodermin, ergokonin, viridian, viridio fungin A, B and C produced by the fungus act in antibiosis [22]. Sharon *et al.* [29] studied the mechanism involved in the attachment and parasitic processes with special emphasis on the important role of the nematode's gelatinous matrix (gm) in direct nematode-fungus interactions, and suggested that carbohydrate-lectin-like interactions might be involved in the attachment of conidia to the nematodes. The authors also found that parasitism was one of the modes of action of *Trichoderma* species against *Meloidogyne javanica. Trichoderma longibrachiatum* produced nematotoxic concentrations of acetic acid. Secondary metabolites from fungi also contained compounds which were toxic to plant parasitic nematodes [17, 30]. *Trichoderma* are also known to produce toxins and antibiotics like malformin, hadacidine, gliotoxin, viridian and penicillin [37], which might contribute to the inactivity and immobility of the nematodes. Parasitic interactions between *Trichoderma* and nematodes might take place in soil, on root surfaces [29] and in the rhizosphere, sites that could be colonized by these opportunistic avirulent plant symbionts [18]. The improved attachment and parasitism observed in vitro could facilitate the development of new strategies to affect interactions between the nematode, plant and fungus for successful biocontrol.

The tested species of *Trichoderma* show a significant effect on the activity of nematodes. The results indicated that *T. harzianum* followed by *T. longibrachiatum, T. viride, T. koningii* and *T. hamatum* were effective in controlling the plant parasitic nematodes *Helicotylenchus* sp. and *Scutellonema* sp. *Trichoderma* sp. was considered imperfect filamentous (Deuetromycetes, Hyphomycetes, Phialasporace, Hyphales, Dematiaceae), and was the most common saprophytic fungi in the rhizosphere found in almost any soil.

Using beneficial fungi for control of plant disease is a useful and acceptable method for farmers. As such, the fungal microbe *Trichoderma* sp. can be used in biological control in the Integrated Pest Management (IPM) programme to achieve good success. Among the tested species of *Trichoderma, Trichoderma harzianum* is a potential candidate. These results are in agreement with [7, 15, 27, 34, 35].

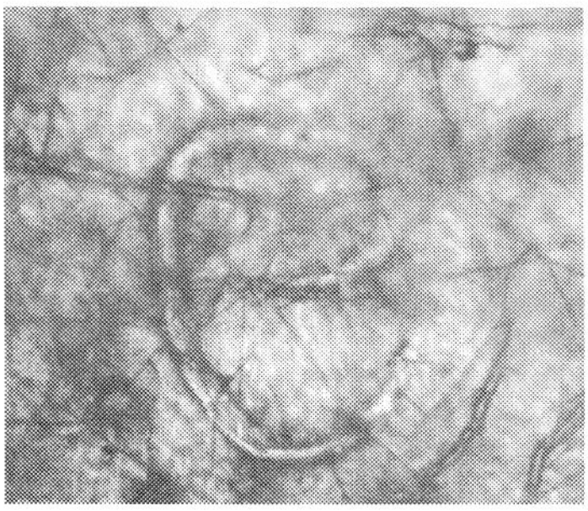

Figure 12. Effect of fungal inoculums of *Aspergillus* sp. on J_2 *M. javanica* (8 days after inoculation) 10 X 10x

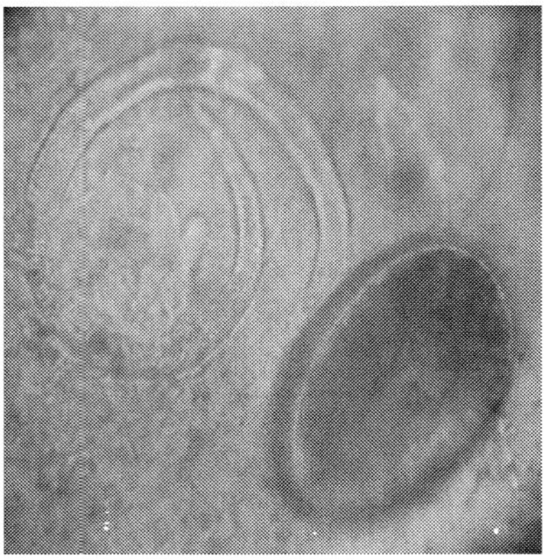

Figure 13. Effect of fungal inoculums of *Aspergillus* sp. on egg and J_2 *M. javanica* (39 days after inoculation) 10 X 10x

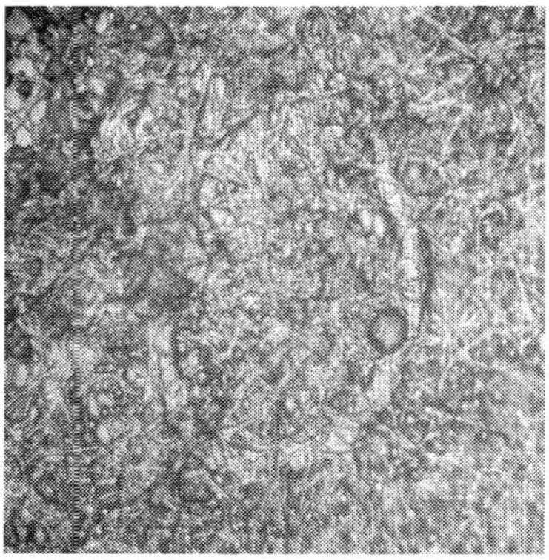

Figure 14. Effect of fungal inoculums of *Mucor* sp. on egg and J_2 *M. javanica* (39 days after inoculation) 10 X 10x

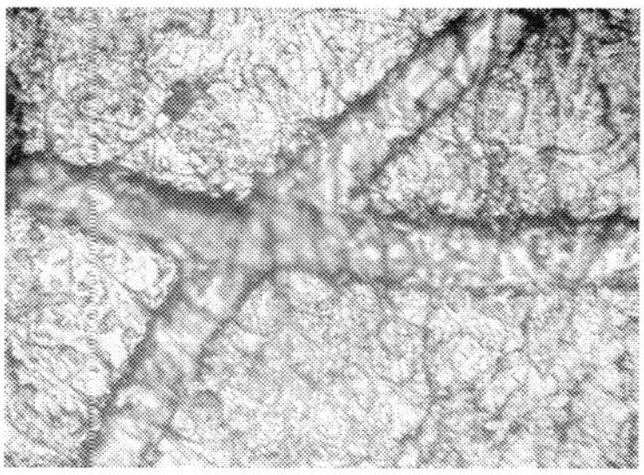

Figure 15. Effect of fungal inoculums of *Mucor* sp. on J_2 *M. javanica* (39 days after inoculation) 10 X 40x

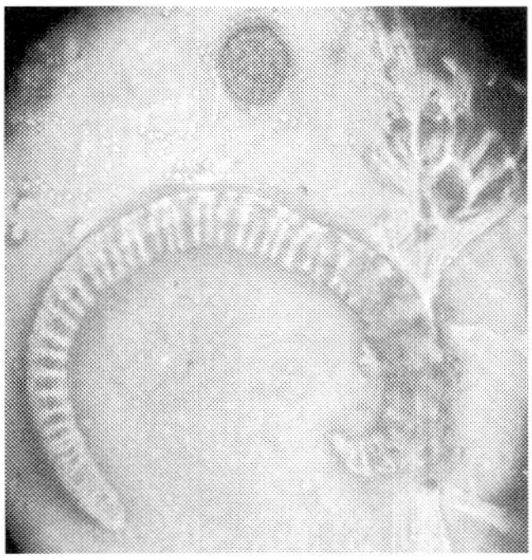

Figure 16. Effect of fungal inoculums of *Paeceilomyces* sp. on egg and J_2 *M.* (39 days after inoculation) 10 X 10x

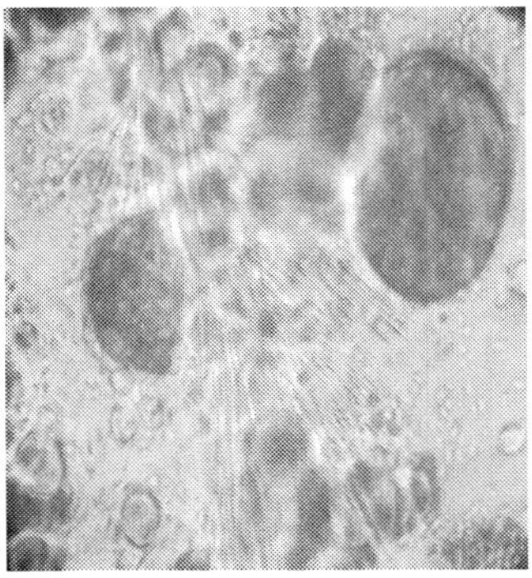

Figure 17. Effect of fungal inoculums of *Paeceilomyces* sp. on egg of *M. javanica javanica* (39 days after inoculation) 10 X 40x

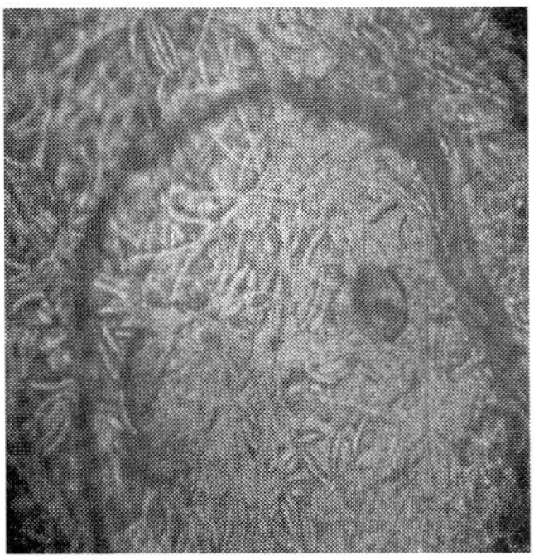

Figure 18. Effect of fungal inoculums of *Penicillium* sp. on egg and J_2 *M.javanica* (39 days after inoculation) 10 X 10x

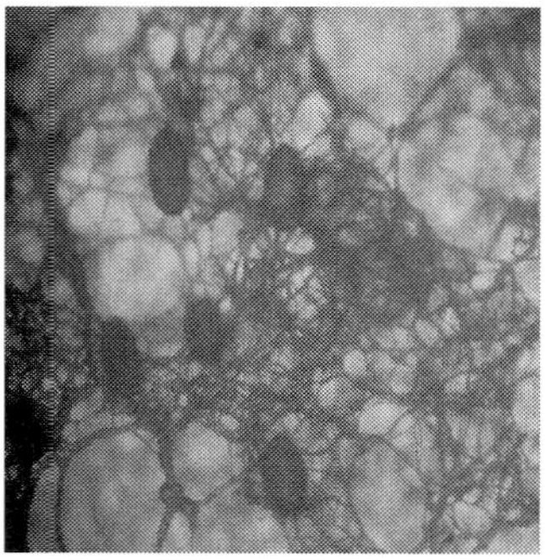

Figure 19. Effect of fungal inoculums of *Penicillium* sp. on egg of *M. javanica* (8 days after inoculation) 10 X 10x

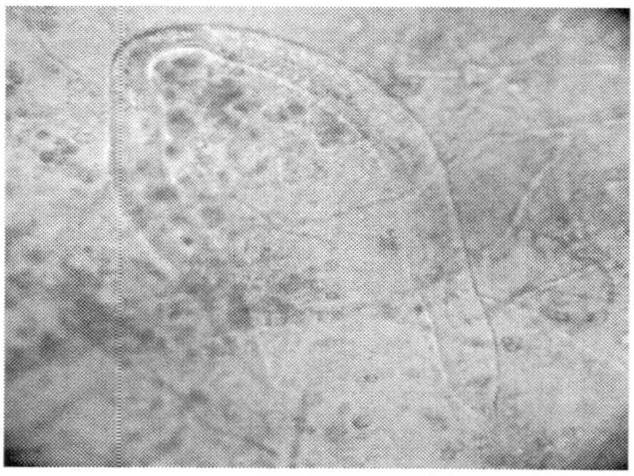

Figure 20. Effect of fungal inoculums of *Trichoderma* sp. on J_2 of *M. javanica* (39 days after inoculation) 10 X 10x

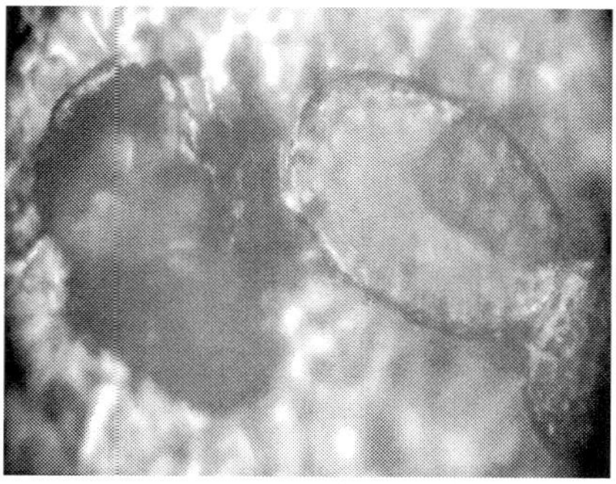

Figure 21. Effect of fungal inoculums of *Trichoderma* sp. on egg of *M. javanica* (39 days after inoculation) 10 X 10x

Figure 22. Effect of different inoculums of J_2*Meloidogyne javanica* on mulberry plants (Var. S_{10})

Figure 23. Galled roots of mulberry plant (Var.S_{10}) due to *M. javanica*

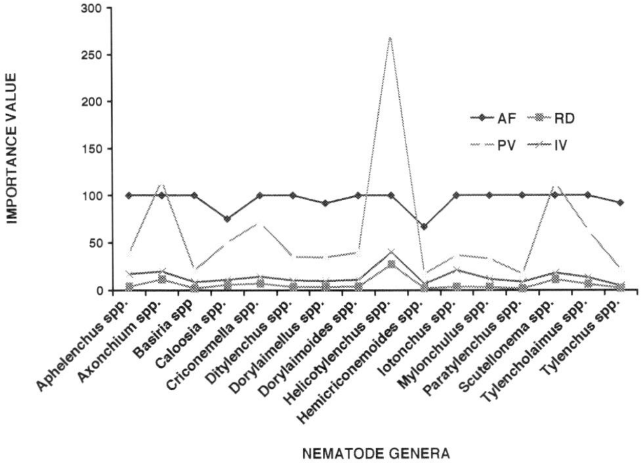

Figure 24. Graph showing absolute frequency, relative density, prominence value and importance value of soil and plant parasitic nematodes associated with mulberry plants at Govt. Silk Farm, Wangbal, Thoubal District, during the year 2006

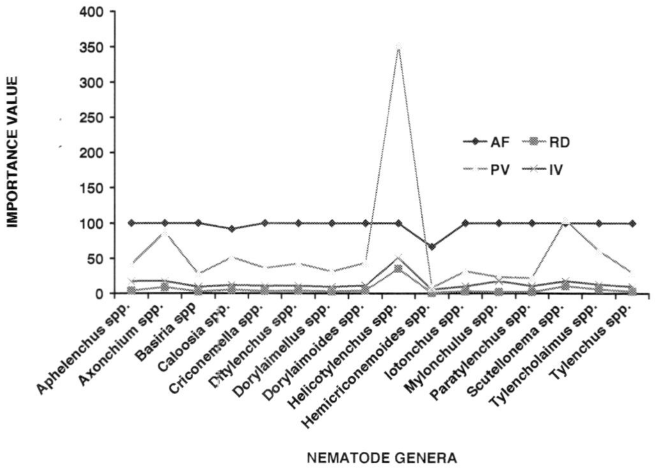

Figure 25. Graph showing absolute frequency, relative density, prominence value and importance value of soil and plant parasitic nematodes associated with mulberry plants at Govt. Silkfarm, Wangbal, Thoubal District during the year 2007

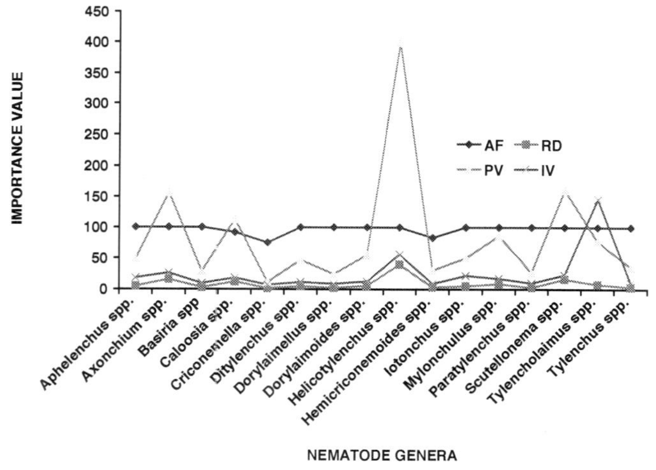

Figure 26. Graph showing absolute frequency, relative density, prominence value and importancevalue of soil and plant parasitic nematodes associated with mulberry plants at Govt. Silk Farm, Wangbal, Thoubal District during the year 2008

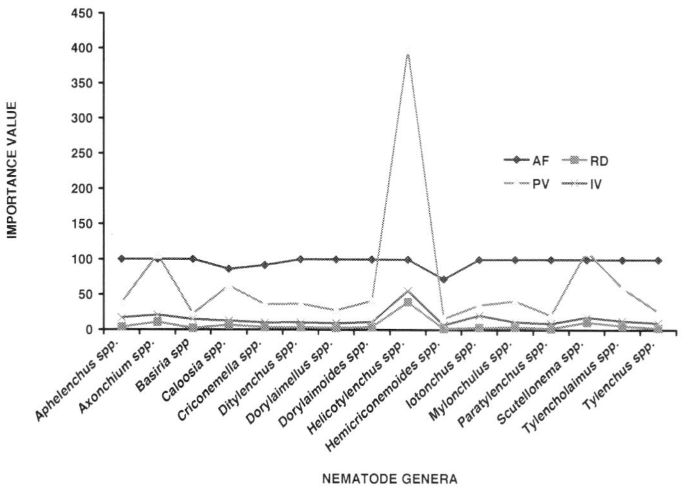

Figure 27. Graph showing absolute frequency, relative density, prominence value and importance value of soil and plant parasitic nematodes associated with mulberry plants at Govt. Silk Farm, Wangbal, Thoubal District during years the 2006–2008

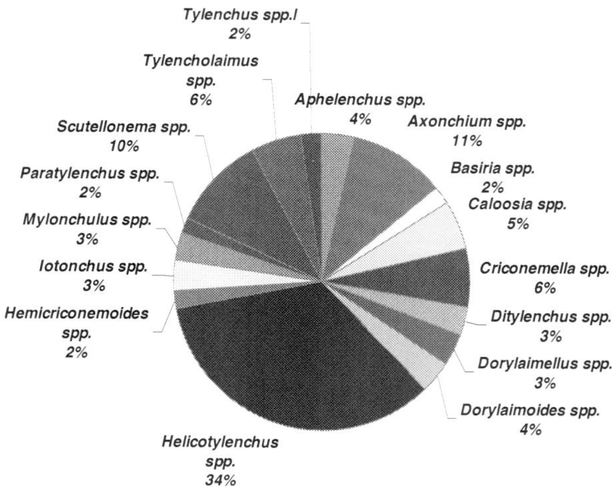

Figure 28. Pie-chart representation of the nematode genera during the year 2006

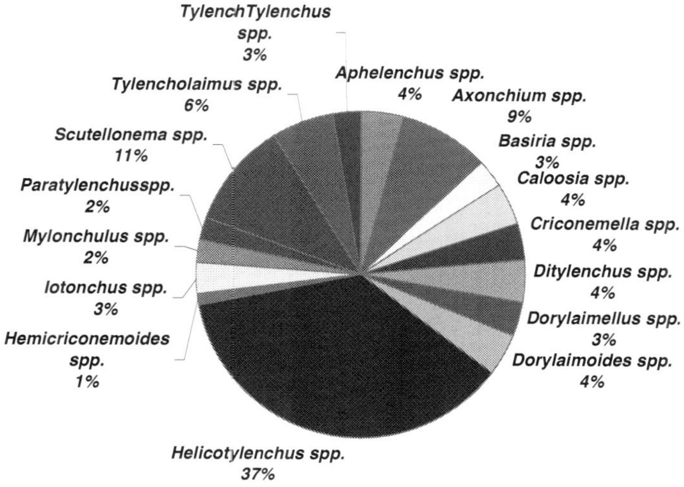

Figure 29. Pie-chart representation of the nematode genera during the year 2007

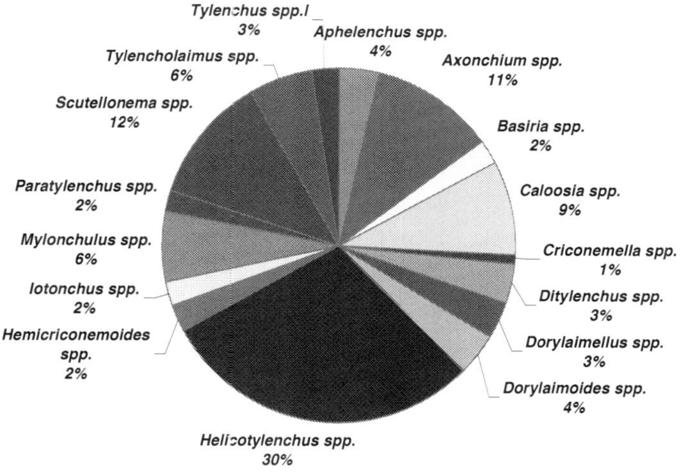

Figure 30. Pie-chart representation of the nematode genera during the year 2008

http://dx.doi.org/10.5772/60082

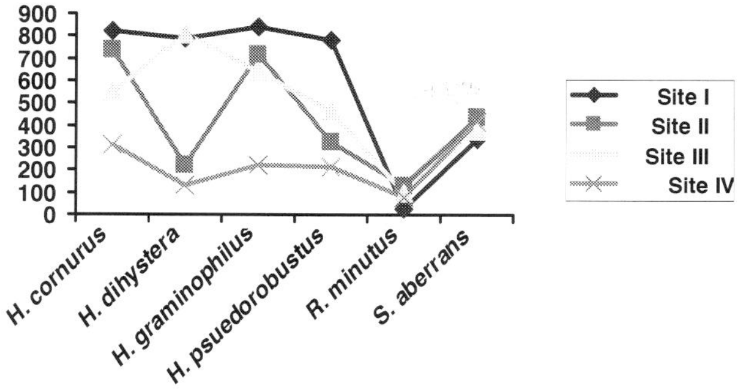

Figure 31. Total nematode population of the family Hoplolaimidae at four different sites in valley districts of Manipur

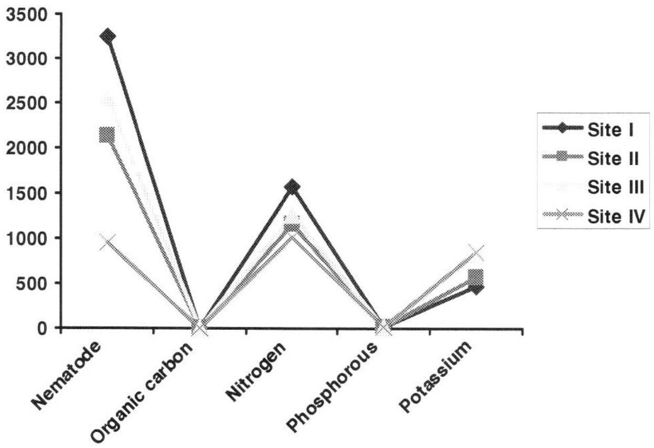

Figure 32. Relationship between the total nematode populations at four sites with their physical-chemical parameters

Author details

L. Bina Chanu, N. Mohilal and M. Manjur Shah*

*Address all correspondence to: mmanjurshah@gmail.com

Department of Life Sciences, Section of Parasitology, Manipur University, Canchipur, Imphal, Manipur, India

References

[1] Abdel-Bari, N. A., Aboul-Eid, H. Z., Anter, E. A. and Noweer, E. A. Effects of different fungal filtrates on *Meloidogyne incognita* larvae in laboratory bioassay tests. *Egyptian J. Agronematol.* 4: 49–69.2000

[2] Add-Elmoity, Riad, F. W. and El-Eraki, S. Effect of single mixture of antagonistic fungi on the control of root-knot nematode *Meloidogyne incognita*. *Egyptian J. Agric.* 71: 91–101.1993

[3] Al-Fattah, A., Dababat, A and Sikora, R. A. Use of *Trichoderma harzianum* and *Trichoderma viride* for the biological control of *Meloidogyne incognita* on tomato. *Jordan J. Agric. Sci.* 3 (3): 297–309.2007

[4] Ali, A. H. H. and Barakat, M. I. E. Utilisation of *Trichoderma harzianum* as a biocontrol agent against root-knot nematode *Meloidogyne incognita*. *Egyptian J. Biol. Pest Control.* 4: 67–77.1994

[5] Badr, S. T. A.: Effects of seven fungal filtrates, singly and combined with three nematicides on *Meloidogyne javanica* juveniles. *Egyptian J. Agronematol.* 5: 105–113.2001

[6] Baker, R. and Griffin, G. J. Molecular strategies for biological control of fungal plant pathogens. In R. Reuveni (ed.), *Novel Approaches to Integrated Pest Management*, pp. 153–82. Boca Raton, FL, USA: Lewis Publisher. 1995

[7] Bokhari, F. M. Efficacy of some *Trichoderma* species in the control of *Rotylenchulus reniformis* and *Meloidogyne javanica*. *Archives Phytopathol. and Plant Prot.* 42 (4): 361–369.2009

[8] Boland, G. J. Biological control of plant diseases with fungal antagonists: Challenges and opportunities. *Canadian J. Plant Pathol.* 12: 295–299.1990

[9] Chet, I. *Trichoderma* - application, mode of action and potential as biocontrol agent of soil borne plant pathogenic fungi. In: I Chet (ed.), *Innovative Approaches to Plant Disease Control*, pp. 137-160. New York: John Wiley & Sons. 1987

[10] Deka, S. B. K. Studies on nematodes associated with mulberry, *Morus alba* L. M.Sc. Thesis, TNAU, Coimbatore, Tamil Nadu, India.1994

[11] Dubey, S. C., Suresh, M. and Singh, B. Evaluation of *Trichoderma* species against *Fusarium oxysporum* fsp. *circeris* for integrated management of chick pea wilt. *Biological Control.* 40: 118–127.2007

[12] Duffy, B. K., Simon, A. and Weller, D. M. Combination of *Trichoderma koningii* with fluorescent pseudomonads for control of take-all on wheat. *Phytopathol.* 86: 188-194.1996

[13] Elad, Y., Zimmand, G., Zags, Y., Zuriel, S. and Chet, I. Use of *Trichoderma harzianum* in combination or alteration with fungicides to control cucumber grey mold (*Botrytis cinerea*) under commercial greenhouse condition. *Plant Pathology.* 42: 324–356.1993

[14] Elad, Y. Biological control of foliar pathogens by means of *Trichoderma harzianum* and potential modes of action. *Crop Prot*. 19: 709–714.2000

[15] Farouk, M. I., Rahman, M. L. and Bari, M.A. Management of root - knot nematode of tomato using *Trichoderma harzianum* and organic soil amendment. *Bangladesh J. Plant Pathology*. 18: 33–37.2002

[16] Freeman, S., Minz, D., Kolesnik, I., Barbul, O., Zreibil, A., Maymon, M., Nitzani, Y., Kirshner, B., Rav-David, D., Bitu, A., Dag, A., Shafir, S. and Elad, Y. *Trichoderma* biocontrol of *Collectotrichum acutatum* and *Botrytis cinerea* and survival in strawberry. *European J. Plant Pathology*. 110: 361–370.2004

[17] Hallmann, J. and Sikora, R. A. Toxicity of fungal endophyte secondary metabolites to plant parasitic nematodes and soil-borne plant pathogenic fungi. *European J. Pl. Pathol*. 102: 155–162.1996

[18] Harman, E.G. (1991): Seed treatments for biological control of plant diseases. *Crop Protection*. 10: 166 – 171.

[19] Inbar, J., Abramsky, M., Cohen, D. and Chet, I. Plant growth enhancement and disease control by *Trichoderma harzianum* in vegetable seedlings grown under commercial conditions. *European J. Plant Pathol*. 100: 337–346.1994

[20] Kerry, B. R. Biological control. In R. H. Brown and B. R. Kerry (eds.), *Principles and Practices of Nematode Control in Crops*, pp. 233–263. New York: Academic Press.1987

[21] Oostendrop, M. and Sikora, R. A. Utilisation of antagonistic rhizobacteria as a seed treatment for the biological control of *Heterodera schachtii* in sugarbeet. *Revue de Nematol*. 12: 77–83.1989

[22] Orole, O. O. and Adejumo, T. O. Activity of fungal endophytes against four maize wilt pathogens. *African J. Microbiol*. 3(1): 969–973.2009

[23] Pandey, G., Pandey, R. K. and Paul, H. Efficacy of different levels of *Trichoderma viride* against root-knot nematode in chickpea (*Cicer arientinum* L.). *Annu Plant Protect*. 26: 971–977.2003

[24] Papavizas, G. C. *Trichoderma* and *Gliocladium* biology, ecology, and potential for biocontrol. *Annual Review Phytopathol*. 23: 23–54.1985

[25] Parvatha, R. P., Rao, M. S. and Nagesh, M. Management of citrus nematode, *Tylenchulus semipenetrans* by integration of *Trichoderma harzianum* with oil cakes. *Nematol. Medit*. 24: 265-267.1996

[26] Rao, V. R. and Swarup, G. Pathogenicity of the spiral nematode, *Helicotylenchus dihystera* to sugarcane. *Indian J. Nematol*. 4 (2):160–166.1974

[27] Reddy, P. P., Rao M. S. and Nagesh, M. Management of the citrus nematode, *Tylenchulus semipenetrans* by integration of *Trichoderma harzianum* with oil cakes. *Nematologia Mediterranea*. 24: 265–267.1996

[28] Sharon, E., Bar-Eyal, M., Chet, I., Herrera-Etrella, A., Kleifeld, O. and Spiegel, Y. Biological control of the root-knot nematode *Meloidogyne javanica* by *Trichoderma harzianum*. *Phytopathology*. 91 (7): 687–693.2001

[29] Sharon, E., Chet, I., Viterbo, A., Bar-Eyal, M., Nagan, H. and Samuels, G. J. Parasitism of *Trichoderma* on *Meloidogyne javanica* and role of the gelatinous matrix. *European J. Plant Pathol*. 118: 247–258.2007

[30] Sikora, R. A., Niere, B. and Kimenju, J. Endophytic microbial biodiversity and plant nematode management in African agriculture. In P. Neuenschwander, C. Borgermeister and J. Langewalder (eds.), *Biological Control in IPM Systems in Africa*, pp. 179–192.2003

[31] Sivan, A. and Chet, I. Microbial control of plant diseases. In R. Mitchell (ed.), *New Concepts in Environmental Microbiology*, pp. 335-354. New York: Wiley-Liss Inc.1992

[32] Spiegel, Y. and Chet, I. Evaluation of *Trichoderma* spp. as a biocontrol agent against soil-borne fungi and plant-parasitic nematodes in Israel. *Integrated Pest Management Reviews*. 3: 169–175.1998

[33] Stephen, Z. A., Hassoon, I. K. and Antoon, B. G. Use of biocontrol agents and nematicides in control of *Meloidogyne javanica* root-knot nematode on tomato and eggplant. *Pakistan J. Nematol*. 16: 151–155.1998

[34] Stephen, Z.A., El-Behadli, A.H., Al-Zahroon, H.H., Antoon, B.G., Georgees, S.S.H. (1996): Control of root – knot - wilt disease complex on tomato plants. *Dirasat Agric. Sci*. 23: 13 – 16.

[35] Stephen, Z.A., Hassan, M.S. and Hasoon, I.K. (2002): Efficacy of fenamiphos, *Trichoderma harzianum*, *Paeceilomyces lilacinus* and some organic soil amendments in the control of root-knot root-rot wilt disease complex of eggplant. *Arabian J. Plant Protection*. 20: 115.

[36] Stirling, G. R. *Biological Control of Plant Parasitic Nematodes*. CAB International, Wallington, UK. 282.1991

[37] Subramanian, C.V. (1964): Predatory observations on host parasitic relationships on plant disease. *Indian Phytopath.Soc.Bull*. 2: 5 – 17.

[38] Thorne, G. *Principles of Nematology*. New York, Toronto & London: McGraw-Hill Book Company.1961

[39] Waksman, S. A. and Fred, B. A tentative outline of the plate method for determining the number of micro-organisms in the soil. *Soil Sciences*. 14: 27–28.1922

[40] Windham, G. L., Windham, M. T. and Williams, W. P. Effects of Trichoderma spp. on maize growth and Meloidogyne arenaria reproduction. Plant Des. 73: 493–494.1989

Section 2

Microbiology and Human Health

Chapter 2

Emergence and Pandemic Potential of Avian Influenza A (H7N9) Virus

Wen Su and Hongxuan He

Additional information is available at the end of the chapter

http://dx.doi.org/10.5772/61007

Abstract

New genotypes of influenza A virus arise quickly and frequently around the world due to continued mutation and reassortment. Novel influenza A viruses pose a direct threat to immunologically naïve humans. A prime example is the recent emergence of avian-origin H7N9 viruses that have become enzootic in China. Three waves of the H7N9 breakout that began in March 2013 have resulted in 571 human cases and over 212 deaths as of 23 February 2015. Real-time influenza surveillance at the wild bird–human interface is essential to limit the outbreak in scale and geographic distribution and to understand the pandemic potential of the H7N9 avian influenza.

Influenza A viruses continue to evolve via several mechanisms, especially by point mutation and reassortment [1], which conduce to the emergence of new strains with epidemic or pandemic potential. In the last 100 years, influenza A viruses have transmitted to and spread among humans, resulting in at least four pandemics [2]. These outbreaks lead to huge economic losses. The recent outbreak of H7N9 viruses following the pandemic spread of 2009 pandemic influenza A(H1N1)pdm09 virus has caused more than 600 human infections, with nearly 30% mortality (http://www.who.int/influenza/human_animal_interface/influenza_h7n9/en/). Multiple distinct lineages of H7N9 viruses have established in chickens [3]. Of note, the increasing epidemiological evidences indicated that H7N9 viruses have limited transmissibility in a family cluster [4]. Therefore, a comprehensive understanding of the mechanisms that determine viral transmissibility, pathogenicity, and evolution is of importance for pandemic preparedness.

Keywords: Influenza, H7N9, evolution, re-assortment

1. Introduction

1.1. Influenza A viruses

Influenza A viruses belong to the family Orthomyxoviridae. The genome of influenza A viruses is comprised of eight segments of single-stranded, negative-sense RNA. The total genome size is about 13,588 bases. There are two viral surface glycoproteins, hemagglutinin (HA) and neuraminidase (NA), in the viral membrane. On the basis of the antigenic variation of the HA and NA, influenza A viruses were classified into different subtypes. Thus far, 18 HA subtypes (H1–H18) and 11 NA subtypes (N1–N11) have been detected in the wild [5]. In theory, influenza A viruses can yield up to 198 combinations. Currently, investigation confirmed that more than nine combinations (H3N2, H1N1, H5N1, H7N7, H7N9, H5N6, H10N8, H2N2, and H9N2) were able to infect humans (Figure 1).

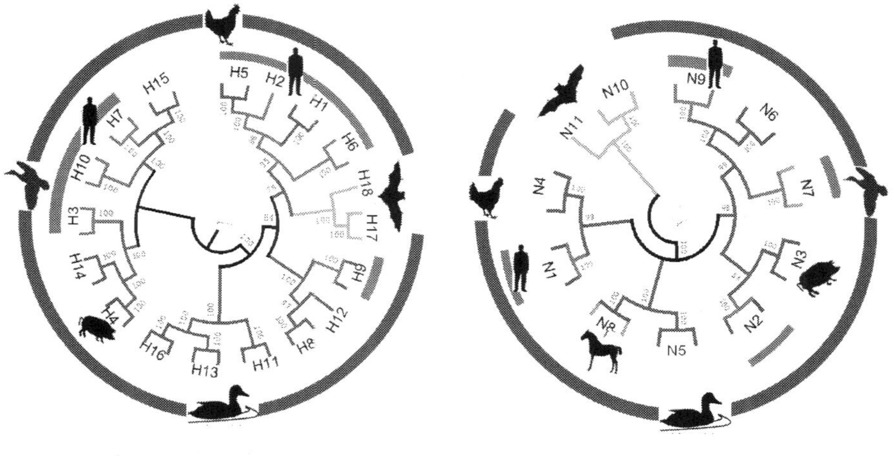

Figure 1. Phylogenetic trees of surface genes of all subtypes of influenza A viruses. The rooted phylogenetic tree was generated based on a full-length nucleotide sequence comparison among influenza A viruses. The representative sequence of each subtype was obtained from the Influenza Virus Resource of NCBI for multiple alignments, and the phylogenetic tree was generated by the ClustalW algorithm in Mega version 6.0 using UPGMA method. The bootstrap values are given at each node. Animal symbols denote the host of influenza A viruses. Subtypes of influenza A viruses have been identified in different hosts, as indicated: blue, birds; red, human; purple, pig; cyan, bat; green, horse.

The genome of influenza A viruses can code for up to 14 proteins [6], which are involved in the influenza virus life cycle [7]. HA mediates the binding to the receptors on host cells and the fusion of the virus and host cell membranes for vRNA release into the cytoplasm. NA is responsible for the efficient release of progeny viruses from infected cells. Replication and transcription and viral RNAs (vRNP) are carried out by the ribonucleoprotein complexes (RNPs) containing three polymerase proteins, basic polymerase 2 (PB2), basic polymerase 1 (PB1), and acidic polymerase (PA), and the nucleocapsid protein (NP). The PB1 also encodes other polypeptides in overlapping open reading frames: PB1-F2 and PB1-N40. PB1-F2 protein, encoded by the +1 reading frame of PB1, functions to kill host immune cells, responding to influenza virus infection [8]. The role of PB1-N40 is associated with PB1 and PB1-F2 [9]. PA gene can code a previously unknown PA-X, which functions to repress cellular gene expression [10]. M1 is the viral matrix structural protein and the ion-channel protein M2 is incorporated in the viral membrane. M42 was identified, which functionally complements M2 [11]. NS gene encodes two proteins: xlinkand NEP. xlink is an interferon antagonist that blocks the activation of transcription factors and NEP is involved in the nuclear export of RNPs into the cytoplasm before virus assembly. The roles of HA and PB2 proteins in viral transmissibility are discussed in more detail later.

2. Highly pathogenic avian influenza viruses

On the basis of the virulence in chickens, influenza A viruses of the H5 and H7 subtypes can be divided into highly pathogenic avian influenza (HPAI) and low-pathogenic avian influenza (LPAI) viruses. In the field, a few H5 and H7 subtype avian influenza viruses are the HPAI phenotype. In general, HPAI viruses are considered to emerge in poultry infected by LPAI viruses. Since 1997, when it was first reported in Hong Kong [12, 13], the highly pathogenic avian influenza A(H5N1) viruses have spread through wild and domestic bird populations across Asia and into Europe, the Middle East, and Africa [14]. Currently, HPAI Asian-origin H5N1 viruses have become endemic in poultry in six countries (Bangladesh, China, Egypt, India, Indonesia, and Vietnam) [15].

H5N1 viruses continue to undergo dynamic evolution [14]. HPAI A(H5N1) viruses have resulted in outbreaks in waterfowl, the natural reservoir of influenza A viruses [16]. Ten years since highly pathogenic Asian-origin avian influenza A (H5N1) viruses began to infect migratory waterfowl in Qinghai Lake in western China in 2005 [16, 17], there is an unprecedented panzootic in more than 63 countries across three continents [18]. They have led to severe economic damage to poultry industry, and have transmitted to humans with lethal consequences. As of 31 March 2015, 826 human infections with 440 deaths have been reported in 16 countries (http://www.who.int/influenza/human_animal_interface/EN_GIP_201503031 cumulativeNumberH5N1cases.pdf?ua=1). Compared with seasonal influenza epidemics and the 2009 H1N1 pandemic, the Asian-origin HPAI H5N1 virus is strikingly severe and fatal, with 52.3% mortality rate. Although several family clusters of H5N1 virus infection have been reported [19], sustained human-to-human infection has not been detected.

Several studies indicated that HPAI Asian-origin H5N1 viruses that are capable of respiratory droplet transmission among mammals can be generated by mutation and reassortment. Herfst et al. introduced two well-known substitutions (Q226L, G228S in H3 HA numbering) in the receptor-binding site (RBS) of HA and a well-known glutamic acid (Glu) to lysine substitution at position 627 (E627K) of PB2 into a human-origin H5N1 isolate [20]. The H5N1 mutant acquired the ability to be transmitted by airborne transmissible among ferrets by further acquiring substitutions after serial passage [20]. Linster et al. stated that two sets of five mutations are identified (E627K in PB2; H99Y in PB1; H103Y, T156A, and either Q222L or

G224S in HA), either of which is sufficient to confer ferret transmissibility on a human H5N1 isolate. Using a combination method, a reassortant H5 HA/H1N1 virus—comprising H5 HA (from an H5N1 virus) with four mutations and the remaining seven gene segments from a 2009 pandemic H1N1 virus—was transmissible between ferrets by respiratory droplet [21]. In contrast, Zhang et al. created many transmissible H5N1 hybrid viruses derived from a lethal H5N1 virus and a 2009 pandemic influenza A(H1N1)pdm09 virus among guinea pigs [22]. However, sustained human-to-human transmission of Asian-origin H5N1 virus has not yet been confirmed in the field.

3. Swine-origin 2009 H1N1 influenza viruses

An outbreak of influenza-like respiratory illness started in a Mexican town in mid-February 2009 [23]. By the end of April, WHO raised repeatedly the influenza alert level from phase 3 to phase 4, and finally, to phase 6 (the pandemic phase) due to the rapid spread of this virus around the world. Most of reported cases outside Mexico and the United States were induced by travelers from both countries. As of May 30, 2010, more than 214 countries across seven continents had reported laboratory confirmed cases of pandemic influenza H1N1 2009, including over 18,138 deaths (http://www.who.int/csr/don/2010_06_04/en/). These viruses had a mortality rate comparable to that of seasonal influenza virus. Currently, influenza A(H1N1)pdm09 continues to circulate as a seasonal virus (http://www.who.int/csr/disease/swineflu/notes/briefing_20100810/en/).

Complete genomic coding sequences indicated influenza A(H1N1)pdm09 viruses emerged from three origins by means of genetic reassortment (i.e., avian/human/swine "triple" reassortants) [24]. These viruses possess avian-origin PB2 and PA genes, a human-origin PB1 gene, and swine-origin HA (H1), NP, NS, NA (N1), and M genes. The human-origin PB1 gene and the avian-origin PB2 and PA genes have been circulating in pigs [7]. However, the factors that contribute to the genesis of influenza A(H1N1)pdm09 viruses remain to be determined.

4. Outbreak of avian-origin H7N9 viruses in China

On 31 March 2013, the national health and family planning commission of China (NHFPCC) first reported human infections caused by a novel avian-origin avian influenza A(H7N9) virus [25]. On April 5, 2013, the Ministry of Agriculture of China first isolated this virus in poultry at an agricultural wholesale market in Shanghai [26]. Since the initial report, three waves of this flu have coursed through the country. Currently, H7N9 viruses have become enzootic in China.

Epidemiological investigations of human cases by the avian influenza A(H7N9) virus showed that most of patients had a history of recent exposure to poultry or a visit to live poultry market, suggesting a close relationship between live poultry market and the spread of flu [27]. The source of avian influenza A(H7N9) virus was immediately traced to live poultry market. After the implementation of compulsory shutdown measures in avian influenza A(H7N9) virus–positive live poultry markets, the government quickly controlled the spread of the virus. However, the elimination of avian influenza A(H7N9) virus is a huge and long-term challenge due to its avirulent nature in poultry.

Phylogeny analysis indicated that the novel avian influenza A (H7N9) virus resulted from the reassortment of recent avian H7N3 viruses and $H_{X(2\ or\ 11)}N9$ influenza viruses with avian influenza A (H9N2) viruses originated from two different groups [28]. As a result, these viruses possess an HA gene of duck-origin H7N3 virus, an NA gene of migratory bird-origin $H_{X(2\ or\ 11)}N9$ virus, and the remaining six viral genes (PB2, PB1, PA, NP, M, and NS) of H9N2 viruses. However, important questions are when, where, and how the avian influenza A (H7N9) viruses are established in poultry and transmitted to humans in China.

The novel avian influenza A(H7N9) virus has several remarkable features. First, this virus has resulted in lethal infections in human and mammalian species, but this virus was low pathogenic for poultry including chickens [29]. Second, this virus possessed a truncated NA protein with a deletion of five amino acids in the stalk, which was associated with viru's virulence [30, 31]. Third, some H7N9 human isolates have a naturally occurring Q226L substitution in the receptor-binding site HA, a pivotal site for the switch of binding receptor [30, 31]. Fourth, PB2 proteins from some human H7N9 isolates have mutations Lys at positions 627 and Asn at 701 [30, 31], whereas the PB2 proteins from bird H7N9 isolates retain Glu at position 627 and Asp at 701. They are important for viral replication at the upper airway of mammalian hosts [32, 33]. These findings suggested that the mutation is positively selected upon replication in the human host.

5. Role of HA in cross-species transmission

HA has a major role in the interspecies transmission of influenza A viruses. The HA of human influenza virus (e.g. influenza A(H1N1)pdm09) preferentially recognizes sialic acid linked to galactose by 6-linkages (Siaa2,6Gal) (human-type receptor), whereas the HA of avian-origin influenza A virus preferentially binds to sialic acid linked to galactose by 2,3-linkages (Siaa2,3Gal) (avian-type receptor). The receptor-binding properties of HA are determined by the amino acid residues in the receptor-binding pocket, which creates microenvironment responsible for viral binding to receptor of host cells. Viruses with leucine (Leu) at position 226 and serine (Ser) at position 228 preferentially bind to human-type receptors, whereas those with glutamine (Gln) and glycine (Gly) at these positions bind to avian-type receptors [34, 35].

Glycan microarray analysis by using a solid-phase binding assay with receptor analogs, revealed that highly pathogenic avian-origin H5N1 virus, in general, specifically binds the avian receptor [34]. When two amino acid substitutions (Q226L and G228S) were introduced in the receptor-binding site (RBS) of H5 HA, highly pathogenic avian-origin H5N1 virus acquired an increased ability to bind to human-type receptors [34], which was due to a cis/trans conformational change in the glycosidic linkage [36]. Several reports have showed that a change of receptor specificity from avian-type receptor to human-type receptor is important for conferring airborne transmission on avian-origin H5N1 virus among mammals [20, 37].

Influenza A(H1N1)pdm09 virus naturally binds the human receptor [38]. This virus is transmissible virus among humans. In the direct glycan receptor-binding assay, the HA of influenza A(H1N1)pdm09 virus exhibited a dose-dependent binding to human-type receptors. Zhang et al. revealed that glutamine at position 226 of HA is a key factor on the transmissibility of influenza A(H1N1)pdm09 virus between mammals [39].

Receptor-binding properties of avian influenza A(H7N9) viruses are diverse. Some H7N9 viruses preferentially bind the avian receptor analog, whereas some have dual receptor-binding property [29, 40-42]. Avian influenza A(H7N9) virus can retain dual receptor properties, although a naturally occurring Q226L of the HA, a pivotal amino acid that mediates receptor-binding specificity was mutated [41], suggesting other factors of avian influenza A(H7N9) virus are probably involved in the receptor-binding specificity. Two substitutions (Q226L and G228S) in the host receptor-binding protein HA contributed to a switch in receptor specificity of an avian H7N9 isolate from dual receptor property to human receptor-binding property (unpublished data). H7N9 influenza virus that binds both avian and human receptor analogs can be transmissible by respiratory droplet among ferrets [29, 40]. However, the transmissibility of this virus is limited and unsustained [4, 40].

6. Role of PB2 in viral transmissibility

The amino acid at position 627 of the PB2 protein is important for airborne transmission of influenza A viruses among mammals. Influenza A viruses with lysine (Lys) at residue 627 of PB2 protein, but not those possessing glutamic acid (Glu) at this position, increased virus replication in mammalian cells at relatively low temperatures [33]. Lys at residue 627 of PB2 is one of five substitutions that proved to be sufficient to confer transmissibility on H5N1 virus [37]. The amino acid at residue 627 of PB2 could increase polymerase activity [37]. In fact, lysine to glutamic acid substitution at position 627 (E627K) of the PB2 could reduce the transmissibility of human influenza viruses [43].

The substitution Glu to Lys in the PB2 proteins at position 627 was positively selected upon replication in human host. In general, the PB2 proteins from human-origin isolates have Lys at position 627, whereas those of viruses isolated from birds retain Glu at this position. The strong selection has been reported previously for influenza A(H7N7), A(H5N1), and novel A(H7N9) infections. For example, the PB2 proteins from most of human H7N9 isolates at positions 627 are Lys. In contrast, the PB2 proteins from H7N9 viruses isolated from birds retain Glu [44].

7. Other proteins in viral transmissibility

The viral ribonucleoprotein (RNP) complex—consisting of the polymerase proteins PB2, PB1, PA, and NP—is one of the most important mechanisms for replication, pathogenticity, and transmissibility of influenza A virus [45]. The levels of RNP activity at 33° and 37°C have an important correlation with virus replication in the upper and lower respiratory tracts of mammalians [32, 33]. Previous studies indicated that high replication efficiency of influenza A virus in the upper respiratory of mammalian host tract contributed to viral transmission between mammalians [32, 33].

Avian-origin H5N1 virus could acquire an essential substitution (His to Tyr) at position 99 of PB1 after serial passage, which was associated with airborne transmission among ferrets [37]. The substitution in PB1 contributed to polymerase transcription and virus replication [20, 37]. Zhang et al. suggested that PA protein and NS protein played a critical role in the airborne transmission between guinea pigs [46]. Therefore, airborne transmission of avian influenza A(H5N1) virus remains a polygenic trait. Transmission of avian influenza A(H7N9) virus also is a polygenic trait [29].

8. Prevention and control

For the treatment of influenza A virus infections, two classes of antiviral drugs—M2 ion channel inhibitors and neuraminidase inhibitors—are at present approved for the control of viral infection. However, the emergence of drug-resistant strains is a challenge. Amantadine and rimantadine are licensed for use against influenza A viruses. However, many strains from multiple subtypes of influenza A virus are now resistant to two available antiviral drugs [7]. Based on a serine to asparagine substitution at position 31 (S31N) of the M2 protein, novel H7N9 viruses are predicted to be resistant to adamantane. Therefore, this drug is not recommended for use [25].

Two neuraminidase inhibitors, Tamiflu (oseltamivir phosphate) and Relenza (zanamivir), are influenza antiviral drugs used by health authorities against recently circulating influenza A viruses, including H7N9 influenza viruses. However, NA inhibitors-resistant influenza viruses are arising in clinical settings. Recently, the rate of oseltamivir-resistant H1N1 viruses, including some H5N1 viruses, is increasing around the world [47-49]. Equally alarming, oseltamivir-resistant H7N9 virus has been reported in a clinical specimen collected two days after commencement of oseltamivir therapy [30]. The H7N9 isolate encoding the NA R292K substitution is highly resistant to oseltamivir and peramivir and partially resistant to zanamivir [50].

In addition to antiviral drugs, inactivated and live attenuated vaccines are available. Traditional influenza vaccines made to protect against seasonal influenza viruses (H1N1 and H3N2 subtypes) are available. In general, vaccines need to be revised and/or developed due to significant antigenic diversity of influenza A viruses. For example, a comparison of novel avian influenza A(H7N9) virus with Eurasian or North American lineages of H7 subtype viruses revealed antigenic differences (http://www.who.int/influenza/vaccines/virus/201302_h5h7h9_vaccinevirusupdate.pdf), suggesting new vaccines need to be developed on the basis of the HA and NA genes of H7N9 candidate vaccine stains. Of note, several new strategies are now in various stages of development. In one example, a "universal" vaccine focuses on regions of viral proteins that are highly conserved across virus subtypes, and provides broad protection against existing and newly emergent unmatched strains [51]. Antihuman C5a antibody treatment markedly reduced acute lung injury (ALI) and systemic inflammation induced by H7N9 virus infection [52].

9. The future

Currently, new strains of influenza A virus are arising quickly and frequently around the world due to continued evolution. In the past two years alone, at least five new subtypes of influenza A virus: A(H7N9), A(H10N8), A(H5N8), A(H6N1), and A(H5N6), have been reported in China. Little knowledge is known about key factors that determine reassortment, virulence, and transmissibility of influenza virus, although much has been learned [1]. Wild birds have served as an ecologic factor favoring emergence and maintenance of avian influenza viruses in natural ecosystems. Therefore, careful influenza surveillance, especially at the wild bird–human interface, remains essential for pandemic preparedness. From a scientific perspective, the opportunity to watch virus evolution in real time may help bridge this gap of knowledge.

Author details

Wen Su and Hongxuan He*

*Address all correspondence to: hehx@ioz.ac.cn

National Research Center for Wildlife Born Diseases, Institute of Zoology, Chinese Academy of Sciences, Beijing, China

References

[1] Webster RG, Bean WJ, Gorman OT, Chambers TM, Kawaoka Y. Evolution and ecology of influenza A viruses. *Microbiol Rev* 1992; 56 (1):152-179.

[2] Sorrell EM, Schrauwen EJ, Linster M, De Graaf M, Herfst S, Fouchier RA. Predicting 'airborne' influenza viruses: (Trans-) Mission impossible? *Curr Opin Virol* 2011; 1 (6): 635-642.

[3] Lam TT, Zhou B, Wang J, et al. Dissemination, divergence and establishment of H7N9 influenza viruses in China. *Nature* 2015. Epub ahead of print. doi:10.1038/nature14348.

[4] Qi X, Qian YH, Bao CJ, et al. Probable person to person transmission of novel avian influenza A (H7N9) virus in Eastern China, 2013: Epidemiological investigation. *BMJ* 2013; 347:f4752.

[5] Yoon SW, Webby RJ, Webster RG. Evolution and Ecology of Influenza A Viruses. Curr Top Microbiol Immunol 2014; 385:359-75.

[6] Bouvier NM, Palese P. The biology of influenza viruses. *Vaccine* 2008; 26 Suppl 4:D49-53.

[7] Neumann G, Noda T, Kawaoka Y. Emergence and pandemic potential of swine-origin H1N1 influenza virus. *Nature* 2009; 459 (7249):931-939.

[8] Chen W, Calvo PA, Malide D, et al. A novel influenza A virus mitochondrial protein that induces cell death. *Nat Med* 2001; 7 (12):1306-1312.

[9] Wise HM, Foeglein A, Sun J, et al. A complicated message: Identification of a novel PB1-related protein translated from influenza A virus segment 2 mRNA. *J Virol* 2009; 83 (16):8021-8031.

[10] Jagger BW, Wise HM, Kash JC, et al. An overlapping protein-coding region in influenza A virus segment 3 modulates the host response. *Science* 2012; 337 (6091):199-204.

[11] Wise HM, Hutchinson EC, Jagger BW, et al. Identification of a novel splice variant form of the influenza A virus M2 ion channel with an antigenically distinct ectodomain. *PLoS Pathog* 2012; 8 (11):e1002998.

[12] Claas ECJ, Osterhaus ADME, van Beek R, et al. Human influenza A H5N1 virus related to a highly pathogenic avian influenza virus. *Lancet* 1998; 351 (9101):472-477.

[13] Subbarao K, Klimov A, Katz J, et al. Characterization of an avian influenza A (H5N1) virus isolated from a child with a fatal respiratory illness. *Science* 1998; 279 (5349): 393-396.

[14] Li KS, Guan Y, Wang J, et al. Genesis of a highly pathogenic and potentially pandemic H5N1 influenza virus in eastern Asia. *Nature* 2004; 430 (6996):209-213.

[15] Ducatez MF, Olinger CM, Owoade AA, et al. Avian flu: Multiple introductions of H5N1 in Nigeria. *Nature* 2006; 442 (7098):37.

[16] Chen H, Smith GJ, Zhang SY, et al. Avian flu: H5N1 virus outbreak in migratory waterfowl. *Nature* 2005; 436 (7048):191-192.

[17] Liu J, Xiao H, Lei F, et al. Highly pathogenic H5N1 influenza virus infection in migratory birds. *Science* 2005; 309 (5738):1206.

[18] Peiris JS, de Jong MD, Guan Y. Avian influenza virus (H5N1): A threat to human health. *Clin Microbiol Rev* 2007; 20 (2):243-267.

[19] Olsen SJ, Ungchusak K, Sovann L, et al. Family clustering of avian influenza A (H5N1). *Emerg Infect Dis* 2005; 11 (11):1799-1801.

[20] Herfst S, Schrauwen EJA, Linster M, et al. Airborne transmission of influenza A/H5N1 virus between ferrets. *Science* 2012; 336 (6088):1534-1541.

[21] Imai M, Watanabe T, Hatta M, et al. Experimental adaptation of an influenza H5 HA confers respiratory droplet transmission to a reassortant H5 HA/H1N1 virus in ferrets. *Nature* 2012; 486 (7403):420-428.

[22] Zhang Y, Zhang Q, Kong H, et al. H5N1 hybrid viruses bearing 2009/H1N1 virus genes transmit in guinea pigs by respiratory droplet. *Science* 2013; 340 (6139): 1459-1463.

[23] Fraser C, Donnelly CA, Cauchemez S, et al. Pandemic potential of a strain of influenza A (H1N1): early findings. *Science* 2009; 324 (5934):1557-1561.

[24] Dawood FS, Jain S, Finelli L, et al. Emergence of a novel swine-origin influenza A (H1N1) virus in humans. *N Engl J Med* 2009; 360 (25):2605-2615.

[25] Gao RB, Cao B, Hu YW, et al. Human infection with a novel avian-origin influenza A (H7N9) virus. *N Engl J Med* 2013; 368 (20):1888-1897.

[26] Shi JZ, Deng GH, Liu PH, et al. Isolation and characterization of H7N9 viruses from live poultry markets-Implication of the source of current H7N9 infection in humans. *Chin Sci Bull* 2013; 58 (16):1857-1863.

[27] Wang CM, Wang J, Su W, et al. Relationship between domestic and wild birds in live poultry market and a novel human H7N9 virus in China. *J Infect Dis* 2014; 209 (1): 34-37.

[28] Lam TTY, Wang J, Shen YY, et al. The genesis and source of the H7N9 influenza viruses causing human infections in China. *Nature* 2013; 502 (7470):241-244.

[29] Zhang Q, Shi J, Deng G, et al. H7N9 influenza viruses are transmissible in ferrets by respiratory droplet. *Science* 2013; 341 (6144):410-414.

[30] Gao R, Cao B, Hu Y, et al. Human infection with a novel avian-origin influenza A (H7N9) virus. *N Engl J Med* 2013; 368 (20):1888-1897.

[31] Liu D, Shi W, Shi Y, et al. Origin and diversity of novel avian influenza A H7N9 viruses causing human infection: Phylogenetic, structural, and coalescent analyses. *Lancet* 2013; 381 (9881):1926-1932.

[32] Massin P, van der Werf S, Naffakh N. Residue 627 of PB2 is a determinant of cold sensitivity in RNA replication of avian influenza viruses. *J Virol* 2001; 75 (11): 5398-5404.

[33] Hatta M, Hatta Y, Kim JH, et al. Growth of H5N1 influenza A viruses in the upper respiratory tracts of mice. *PLoS Pathog* 2007; 3 (10):1374-1379.

[34] Stevens J, Blixt O, Tumpey TM, Taubenberger JK, Paulson JC, Wilson IA. Structure and receptor specificity of the hemagglutinin from an H5N1 influenza virus. *Science* 2006; 312 (5772):404-410.

[35] Stevens J, Corper AL, Basler CF, Taubenberger JK, Palese P, Wilson IA. Structure of the uncleaved human H1 hemagglutinin from the extinct 1918 influenza virus. *Science* 2004; 303 (5665):1866-1870.

[36] Zhang W, Shi Y, Lu XS, Shu YL, Qi JX, Gao GF. An airborne transmissible avian influenza H5 hemagglutinin seen at the atomic level. *Science* 2013; 340 (6139):1463-1467.

[37] Linster M, Boheemen S, de Graaf M, et al. Identification, characterization, and natural selection of mutations driving airborne transmission of A/H5N1 virus. *Cell* 2014; 157 (2):329-339.

[38] Childs RA, Palma AS, Wharton S, et al. Receptor-binding specificity of pandemic influenza A (H1N1) 2009 virus determined by carbohydrate microarray. *Nat Biotechnol* 2009; 27 (9):797-799.

[39] Zhang Y, Zhang Q, Gao Y, et al. Key molecular factors in hemagglutinin and PB2 contribute to efficient transmission of the 2009 H1N1 pandemic influenza virus. *J Virol* 2012; 86 (18):9666-9674.

[40] Richard M, Schrauwen EJA, de Graaf M, et al. Limited airborne transmission of H7N9 influenza A virus between ferrets. *Nature* 2013; 501 (7468):560-563.

[41] Shi Y, Zhang W, Wang F, et al. Structures and receptor binding of hemagglutinins from human-infecting H7N9 influenza viruses. *Science* 2013; 342 (6155):243-247.

[42] Xiong XL, Martin SR, Haire LF, et al. Receptor binding by an H7N9 influenza virus from humans. *Nature* 2013; 499 (7459):496-499.

[43] Steel J, Lowen AC, Mubareka S, Palese P. Transmission of influenza virus in a mammalian host is increased by PB2 amino acids 627K or 627E/701N. *PLoS Pathog* 2009; 5 (1):e1000252.

[44] Organization WH. Overview of the emergence and characteristics of the avian influenza A (H7N9) virus. Available at: http://www.who.int/influenza/human_animal_interface/influenza_h7n9/WHO_H7N9_review_31May13.pdf?ua=1. (Accessed 2 June 2013).

[45] Gabriel G, Dauber B, Wolff T, Planz O, Klenk HD, Stech J. The viral polymerase mediates adaptation of an avian influenza virus to a mammalian host. *Proc Natl Acad Sci USA* 2005; 102 (51):18590-18595.

[46] Zhang Y, Zhang QY, Kong HH, et al. H5N1 hybrid viruses bearing 2009/H1N1 virus genes transmit in Guinea pigs by respiratory droplet. *Science* 2013; 340 (6139): 1459-1463.

[47] de Jong MD, Tran TT, Truong HK, et al. Oseltamivir resistance during treatment of influenza A (H5N1) infection. *N Engl J Med* 2005; 353 (25):2667-2672.

[48] Le QM, Kiso M, Someya K, et al. Avian flu: Isolation of drug-resistant H5N1 virus. *Nature* 2005; 437 (7062):1108.

[49] Poland GA, Jacobson RM, Ovsyannikova IG. Influenza virus resistance to antiviral agents: A plea for rational use. *Clin Infec Dis* 2009; 48 (9):1254-1256.

[50] Hai R, Schmolke M, Leyva-Grado VH, et al. Influenza A(H7N9) virus gains neuraminidase inhibitor resistance without loss of in vivo virulence or transmissibility. *Nat Commun* 2013; 4:2854.

[51] Schotsaert M, De Filette M, Fiers W, Saelens X. Universal M2 ectodomain-based influenza A vaccines: Preclinical and clinical developments. *Expert Rev Vaccines* 2009; 8 (4):499-508.

[52] Sun S, Zhao G, Liu C, et al. Treatment with anti-C5a antibody improves the outcome of H7N9 virus infection in African green monkeys. *Clin Infect Dis* 2015; 60 (4):586-595.

Chapter 3

Negative and Positive Staining in Transmission Electron Microscopy for Virus Diagnosis

Debora Ferreira Barreto-Vieira and
Ortrud Monika Barth

Additional information is available at the end of the chapter

http://dx.doi.org/10.5772/60511

Abstract

Visualization of virus particles and morphological features depends on the resolution of microscopes. Transmission electron microscopy (TEM) is the starting point for obtaining the best resolution of images. Two different techniques are available and described in this paper. Firstly, negative staining of viral suspensions provides detailed information of virus particles' structure. It is a technique that can be quickly performed and is able to accommodate the highest magnifications of virus particles. Secondly, ultra-thin sections of virus-infected tissues or cell cultures, combined with a positive staining technique can provide information regarding the localization of viruses inside or around cells. These two complementary techniques for investigating the structure of a virus and its parasitic life cycle are presented in this paper.

Keywords: transmission electron microscopy, negative staining, positive staining

1. Introduction

High resolution microscopes are essential for visualizing virus particles. A transmission electron microscope (TEM) fulfils this requirement and is widely used by scientists. Unknown agents of diseases may be quickly detected by using such a microscope. A recent review on the value of electron microscopy and virus structures was presented by Zhang et al. [1].

Shortly prior to the recent recognition of mimiviruses [2], orthopoxviruses had been one of the largest virus particles, in addition to the thin and very long filoviruses. Groups of orthopoxvirus particles can be detected when inside infected cells using the highest magnification available in light microscopy. Nevertheless, an isolated virus particle of ca. 250 nm diameter remains below the resolution of standard light microscopes. With the advent of TEMs and high magnifications of over 100 000x, the structural details of virus particles could then begin to be recognized.

The first demonstration of a viral particle using a TEM was in the form of a member of the orthopoxvirus genus [3]. Following the introduction of the negative staining technique in the 1960s [4], a variety of viruses has been discovered and classified based on morphological characteristics such as symmetry, the presence or absence of the envelope or spikes and the size of these projections [5,6,7,8]. A large number of viruses of medical importance that had formerly never been described such as adenovirus, enterovirus, orthomyxovirus, reovirus and paramyxovirus were identified by TEM after isolation in cell cultures inoculated with clinical specimens [9]. Other viruses like hepatitis B and hepatitis A were detected directly by TEM in samples such as plasma and faeces, following the failure of attempting to try and isolate them in cell cultures. The initial classification of many agents was therefore based on a combination of morphology and genome structure.

Currently, more than 30 000 different viruses comprising 56 families have been identified using TEM and humans have been found to play host to 21 of the 26 families specific for vertebrates [10]. The development of other techniques, e.g., immunofluorescence, enzyme-linked immunosorbent assays (ELISA) and biological molecular methods such as polymerase chain reaction (PCR) have progressively reduced the importance of TEM in virus diagnosis and microbiology. However, compared with other diagnostic methods, TEM still benefits from its rapidity and "open view", i.e., the capability of detecting all pathogens present in a clinical specimen [11]. Therefore, TEM should be utilized as a frontline method in infectious disease emergencies and/ or in suspected cases of bioterrorism [12]. Electron microscope studies were critical in identifying the aetiologic agent of severe acute respiratory syndrome (SARS), a coronavirus, during its 2002/2003 global outbreak [13].

Several methods for specimen preparation have been developed. These can be summarized into two procedures: the negative staining of vesicular content and viral suspensions, and the ultra-thin sectioning of infected tissues and cells. Both techniques have been carried out with complementary results.

2. Negative staining technique

The classic processing of biological specimens observed in a TEM needs fixation, dehydration, sectioning and a selective "staining" of cell and tissue structures. "Staining", a means of receiving coloured images, cannot be effectively used in conjunction with an electron microscope. Instead, the enhancement of structures for TEM observation is effected, usually by impregnation with heavy metal salts of plumb, tungstenium and uranium.

Some biological elements are very small and as a result, sectioning reveals aspects of its content but not about its global surface structure. The result is a bi-dimensional image. Nevertheless, when observed as a whole using the negative staining technique, these elements reveal a tri-dimensional image.

To prepare small biological specimens such as bacteria, viruses, phages, micoplasma, filaments and cell membranes, or even nucleic acids and protein filaments for TEM observation, the special technique of "negative contrast" was developed [4]. Instead of applying a more or less strong positive contrast of structures ("staining"), in this instance, contrast is not applied to the object but to its environment, using heavy metal salts. The electron beam can cross biological material easier than the surrounding space. The result resembles an inverted traditional TEM image (Figures 2, 4-7).The standard staining solution used today is an aqueous 1% phosphotungstic acid with pH 7.2.

Essentially, a fixed or unfixed drop of viral suspension is applied onto a formvar- or collodium-covered electron microscope grid for a few seconds; the liquid is absorbed by a filter paper, then a drop of the staining solution is applied and few seconds later also absorbed (Figure 1). After drying, the specimen is ready to be introduced into the electron beam.

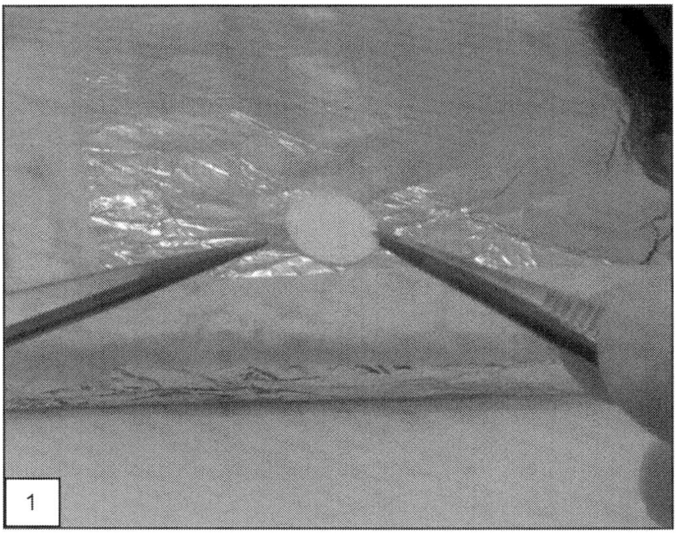

Figure 1. Negative staining procedure. Forceps, grid and white filter paper are the primary tools.

This unbelievably simple technique, at the outset rejected by the scientific community, eventually became a revolutionary approach for studying primarily viral morphology as emphasized by viral diagnostics [14]. Extremely detailed images were obtained and published, revealing substructures and macromolecules as viral antibodies and virus particle spines [5, 9, 15, 16, 17, 18, 19, 20].

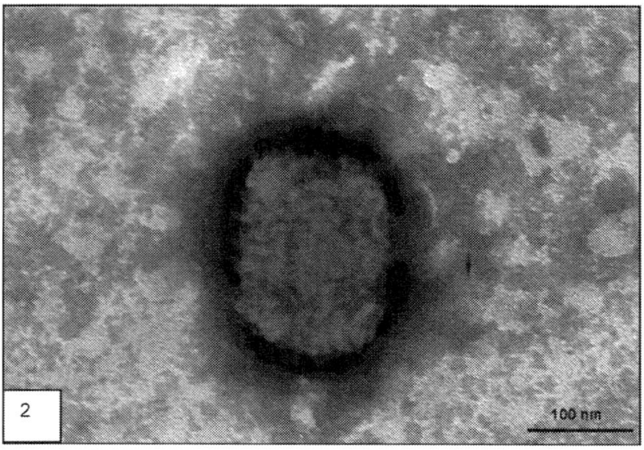

Figure 2. Orthopoxvirus particle, Brazilian Cantagalo strain (negative staining). This virus particle was isolated from a milker's hand nodule, macerated using a drop of distilled water and applied onto a collodium-covered EM-copper grid, then stained using a drop of 2% phosphotungstic acid in distilled water.

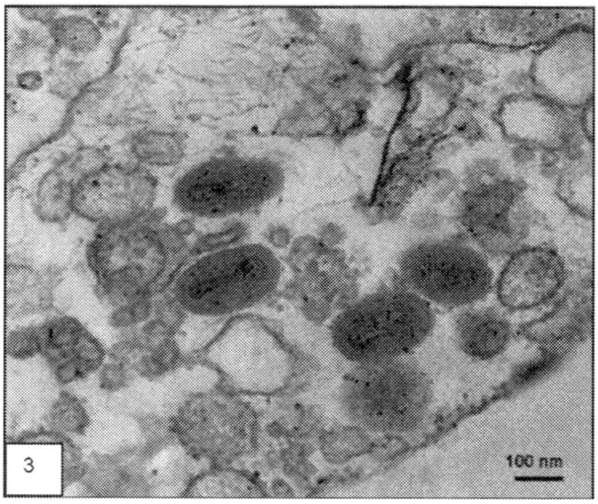

Figure 3. Ultra-thin section of a Vero cell infected with the same strain of virus demonstrated in Figure 2 (positive staining).

In order to gain additional information about small structured elements, several technical steps in sample processing were developed such as immune-electron microscopy (Figure 5), solid-phase-electron microscopy, ammonium-sulphate precipitation, gradient fraction contrasting and particle concentration by diffusion in an agarose layer [21].

Contrasting solutions show a large spectrum of possibilities. Selection must be in accordance with the pH and electrical charge of the sample, of the contrasting solution and of the EM grid

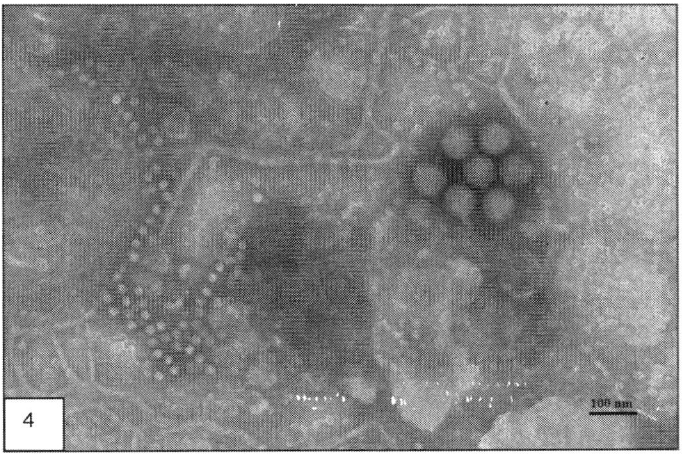

Figure 4. Group of seven adenovirus particles and numerous small adeno-associated virus particles. A drop of clarified faeces from a patient suffering from gastroenteritis was processed as described in Figure 1.

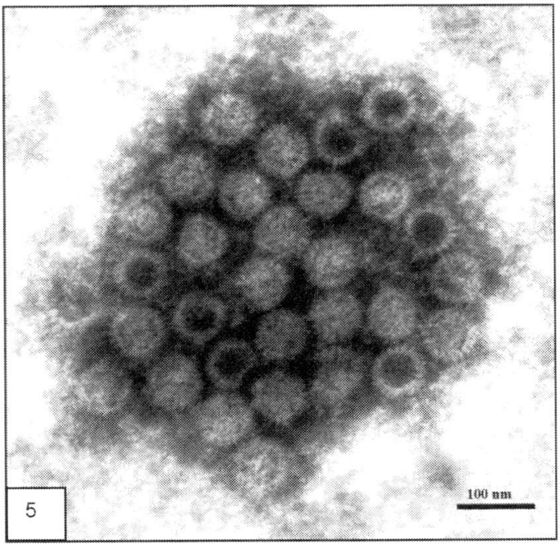

Figure 5. A group of rotavirus particles. Clarified faeces from a patient suffering from gastroenteritis was incubated with a specific antibody prior to being submitted to the negative staining procedure and processed as described in Figure 1. Both several empty and RNA-filled virus particle can be distinguished.

or support [19, 22]. Resulting precipitates and a lack of spreading of the sample are the most common inconveniences. For better spreading and adsorption of virus particles onto a formvar-carbon coated grid, polylysine (poly-L-lysine) is currently being used in our laboratory (Figures 6 to 7).

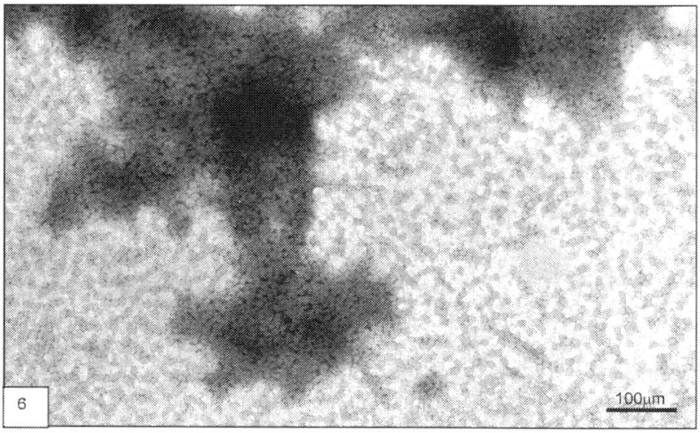

Figure 6. Hepatitis B antigen. One drop of a gradient virus suspension was applied directly onto an electron microscope grid and in sequence stained with 1% alcoholic uranyl acetate. Virus particles are joined together into very dark groups.

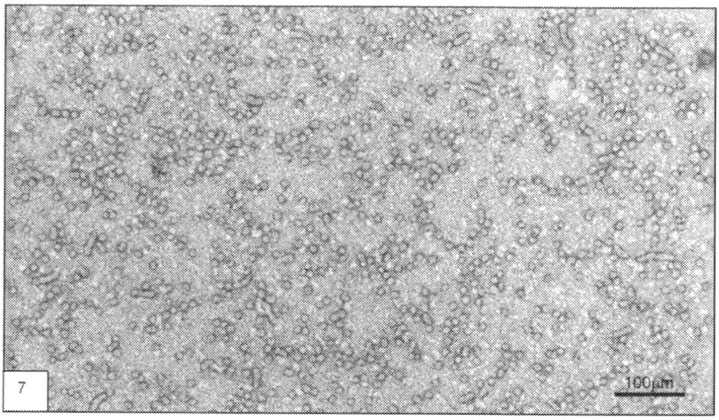

Figure 7. Hepatitis B antigen. One drop of the same virus suspension as shown in Figure 6 was applied to an electron microscope grid formerly coated with polylysine and in sequence stained with 1% alcoholic uranyl acetate. Virus particles are spread all over the grid.

An overview of the negative staining technique development and application was presented by Biel & Gelderblom [6], Harris et al. [23] and more recently, by Schramlová et al. [24]. Benefits or deception are always surprising factors when the electron beam reveals a TEM image inside a dark room.

3. Positive staining technique

The positive staining technique has been used since the late 50s and the early 60s for enhancing the contrast of biological samples (tissues and cell structures, viruses, etc.). Using this technique, as well as negative staining, the samples are incubated in heavy metal salt solutions that react with cellular structures. Uranyl acetate [25] and lead citrate [26] are the most commonly used salts today. Grids containing ultra-thin sections of a sample are incubated for 15 minutes in uranyl acetate; this procedure should be performed in an environment protected from light. Following on, the grids are washed in distilled water and incubated in lead citrate at four to five minutes in an environment free of CO_2. NaOH tablets are used to keep the environment free from reacting with CO_2 (Figure 8). At the end of the procedure the grids are washed in distilled water, air dried and observed with a TEM [27, 28].

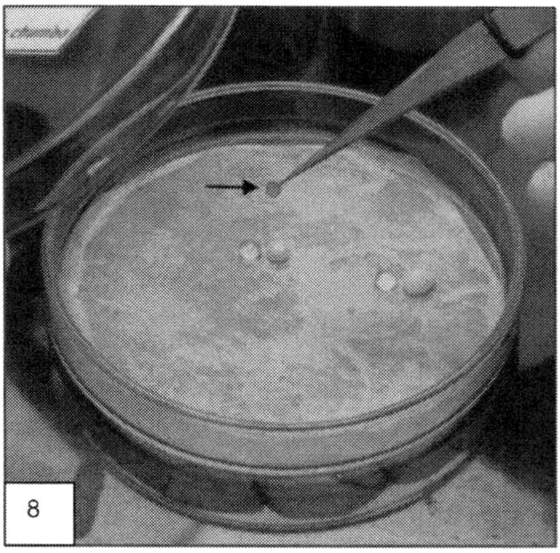

Figure 8. Positive staining. The grid held with a forceps contains ultra-thin sections of a sample (arrow). In the Petri dish, there are two drops of the staining solution beside two tablets of NaOH (Barreto-Vieira et al. 2010).

3.1. Specimen preparation

Cells. Monolayers were fixed in 1% glutaraldehyde in a cacodylate buffer (0.2 M, pH 7.2) and post-fixed with 1% buffered osmium tetroxide, dehydrated in acetone, embedded in epoxy resin and polymerized at 60°C for three days [28, 29]. The blocks were cut into ultra-thin sections 50 to 70 nm thick using a diamond knife adapted to an ultramicrotome (Figure 9). The sections were placed onto copper grids and stained with uranyl acetate and lead citrate and observed in a TEM (Figure 10).

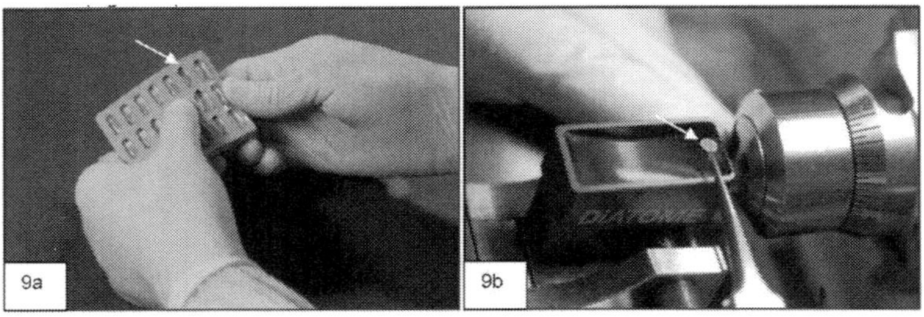

Figure 9. Ultramicrotomy: **a)** silicone support filled with resin blocks (arrow) containing fragments of tissues; **b)** ultrathin sectioning using a diamond knife and a copper grid (arrow) to collect sections.

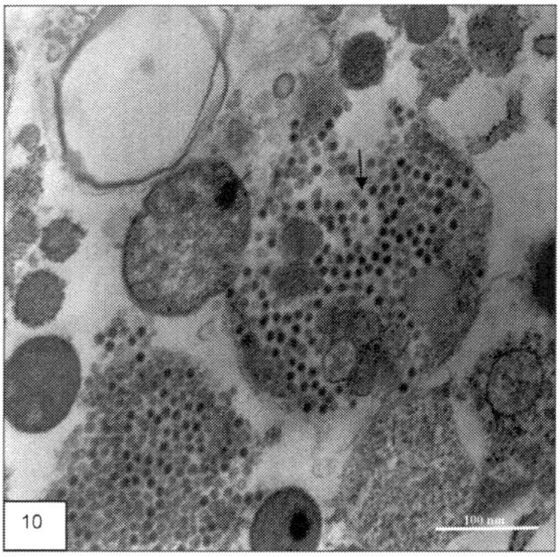

Figure 10. Dengue virus particles (arrow) isolated in the *Aedes albopictus* mosquito cell line (C6/36) from the serum of infected BALB/c mice. Positive staining using lead citrate (arrow) emphasizes viral RNA, while cytoplasmic proteins are emphasized using uranyl acetate.

Fragments of tissues. Glutaraldehyde at 2% in a cacodylate buffer (0.2M, pH 7.2) was routinely used to fix tissue fragments, then post-fixed by immersion in a 1% buffered osmium tetroxide, dehydrated in acetone, embedded in epoxy resin and polymerized at 60°C for three days [28, 29]. Semi-thin 0.5 μm thick sections were cut using a diamond knife. These sections were stained with a methylene blue/azure II solution [30] and observed using a photonic microscope. Ultra-thin sections were placed onto copper grids and stained with uranyl acetate and lead citrate and observed in a TEM.

Tissue fragments from experimentally infected animals. In our experiments, the infected animals (mice) were peritoneally anaesthetized and fixed by perfusion with a 4% paraformaldeyde in a sodium phosphate buffer (0.2M, pH 7.2) for 30 min. The tissues were carefully collected in sequence and the fragments post-fixed by immersion in 2% glutaraldehyde in cacodylate buffer (0.2M, pH 7.2), post-fixed with 1% buffered osmium tetroxide, dehydrated in acetone, embedded in epoxy resin and polymerized at 60°C for three days [28, 29]. Ultra-thin sections were obtained and processed as described above (Figure 11).

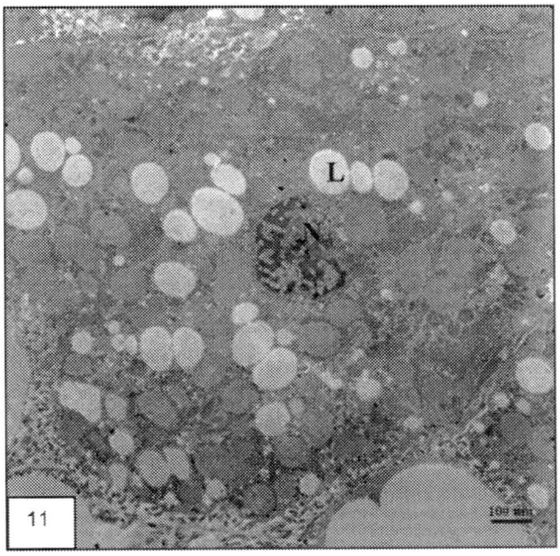

Figure 11. Hepatic tissue of BALB/c mice infected with the dengue virus. Note the numerous lipidic inclusions (L) and a pyknotic nucleus (N). Positive staining emphasizes the chromatin of the hepatocyte nucleus using lead citrate and cytoplasmic proteins using uranyl acetate.

4. Advantages and disadvantages of the TEM

Virus diagnosis via TEM, when compared to other techniques such as immunofluorescence, PCR and ELISA have both advantages and disadvantages.

Advantages: **1)** TEM is the only technique that allows for direct visualization of an aetiologic agent of very small diameter, e.g., viruses; all other diagnostic techniques show indirect results; **2)** samples can be processed and analysed quickly. Following simple and rapid negative staining, TEM "open view" allows for rapid morphological identification and a differential diagnosis of various infectious agents present in a specimen. Likely, intact virions observed with TEM may be indicative of an infectious virus, whereas the detection of antigens or nucleic acid may not always indicate the presence of viable infectious virus particles [31]; **3)** TEM is valuable for differential diagnoses, for example, in patients with vesicular dermatitis, so as to exclude smallpox; **4)** TEM assists in elucidating unknown aetiologic agents in outbreaks, epidemics or pandemics, since it is not necessary to use specific reagents. The causative agents of the recent SARS pandemic and the outbreak of human monkey pox in the USA were quickly identified using TEM [13].

*Disadvantages:***1)** TEM is of low diagnostic sensitivity compared to other diagnostic methods. It requires a high concentration of virus particles (>10^5 particles per mL), both in suspensions and in tissues or cells; **2)** the evaluation of samples cannot be automated with TEM and should be performed by an experienced professional; **3)** TEM is not an appropriate method for screening a large number of samples as in a viral outbreak, due to the high cost of TEM maintenance and the availability of trained personal.

Acknowledgements

Financial support: research fellowship of the Conselho Nacional de Desenvolvimento Científico e Tecnológico (CNPq), process numbers 301525/2009-9 and 304067/2014-18 (second author).

Author details

Debora Ferreira Barreto-Vieira * and Ortrud Monika Barth

*Address all correspondence to: barreto@ioc.fiocruz.br

Laboratory of Morphology and Virus Morphogenesis, Instituto Oswaldo Cruz, Fundação Oswaldo Cruz, Rio de Janeiro, Brazil

References

[1] Zhang Y, Hung T, Song J D, He JS. Electron microscopy: essentials for viral structure, morphogenesis and rapid diagnosis. Science China Life Sciences 2013;56: 421-430.

[2] Saadi H, Pagnier I, Colson P, Cherif J K, Beji M, Boughalmi M, Azza S, Armstrong N, Robert C, Fournous G, La Scola B, Raoult D. First Isolation of *Mimivirus* in a patient with pneumonia. Clinical Infectious Disease 2013;57: e127-134.

[3] Von Borries B, Ruska E, Ruska H. Bakterien und Virus in übermikroskopischer Aufnahme, Klinische Wochenschrift 1938;*17*: 921-925.

[4] Brenner S, Horne RW. A negative staining method for high resolution electron microscopy of viruses. Biochimica et Biophysica Acta 1959;34:103.

[5] Hayat MA, Miller SE. Negative Staining. McGraw Hill Publishing Company, NY, 253p; 1990.

[6] Biel SS, Gelderblom HR. Diagnostic electron microscopy is still a timely and rewarding method. Journal of Clinical Virology 1999; 13: 105-119.

[7] Flewett TH, Bryden AS, Davies H. Virus particles in gastroenteritis. Lancet 1973;2:1497.

[8] Tyrrell DAJ, Almeida J. Direct electron-microscopy of organ culture for the detection and characterization of viruses. Archive of Virology 1967;22: 417-425.

[9] Madeley CR, Field AM. Virus Morphology, 2nd ed., Churchill Livinstone. 295p;1988.

[10] Van Regenmortel MHV, Fauquet CM, Bishop DHL, Carsten EB, Estes MK et al. Virus taxonomy: classification and nomenclature of viruses, 7th Report of the International Committee on Taxonomy of Viruses, Academic Press, San Diego. 1167 p; 2000.

[11] Gentile M, Gelderblom HR. Rapid viral diagnosis: role of electron microscopy. New Microbiology 2005;28: 1-12.

[12] 12] Schatzmayr HG, Barth OM. Bioterrorismo e microorganismos patogênicos. História, Ciências, Saúde – Manguinhos 20: 1735-1749;2013.

[13] Goldsmith CS, Tatti KM, Ksiazek TG, Rollin PE, Comer JA, Lee WW, Rota PA, Bankamp B, Bellini WJ, Zaki SR. Ultrastructural characterization of SARS coronavirus. Emerging Infectious Disease 2004;10: 320-326.

[14] Gelderblom HR. Electron microscopy in diagnostic Virology. Bioforum 2001;5: 64-67.

[15] Horne RW. The examination of small particles. Negative staining methods. In DH KAT, Techniques for Electron Microscopy, Blackwell Scientific Publications, Oxford, 2nd ed, p. 311-355;1965.

[16] Nermut MV. Methoden der Negativkontrastierung. *Meth Samml EM Wiss Verl*, Stuttgart, 3.1.2.3, 22p;1973.

[17] Almeida JD, Atanasiu P, Bradley DW, Gardner PS, Maynard J, Schuurs AW, Voller A, Yolken RH. Manual for rapid laboratory viral diagnosis. World Health Organization Publication 1979;47: 7-10.

[18] Doane FW, Anderson N. Electron Microscopy in Diagnostic Virology. Cambridge University Press, Cambridge. p.178;1987.

[19] Barth OM. Estudos sobre a contrastação negativa de suspensões virais. Revista Brasileira de Biologia 1984; 44: 71-80.

[20] Barth OM, Schatzmayr HG. Classificação e características gerais dos vírus patogênicos para o homem. In JR Coura, *Dinâmica das Doenças Infecciosas e Parasitárias*. Segunda edição, Editora Guanabara Koogan, Rio de Janeiro, p.1680-1694; 2013.

[21] Barth OM. Contrastação Negativa. In W de Souza, *Manual Sobre Técnicas Básicas em Microscopia Eletrônica*. Vol. I. Técnicas Básicas. Sociedade Brasileira de Microscopia Eletrônica, Rio de Janeiro, p.60-73;2007.

[22] Barth OM. The use of polylysine during negative staining of viral suspensions. Journal of Virological Methods 1985; 11: 23-27.

[23] Harris JR, Bhella D, Adrian M. Recent developments in negative staining for transmission electron microscopy. The Americas Microscopy and Analysis 2006;78: 5-9.

[24] Schramlová J, Arientová S, Hulínská D. The role of electron microscopy in the rapid diagnosis of viral infections – review. Folia Microbiology 2010;55: 88-101.

[25] Watson ML. Staining of tissue sections for electron microscopy with heavy metals. II. Application of solutions containing lead and barium. The Journal of Biophysical and Biochemical and Cytology 1958;4: 727-730.

[26] Reynolds. The use of lead citrate at high pH as an electron opaque stain in electron microscopy. Journal of Cell Biology 1963;17: 208-212.

[27] Attias M. Desidratação, inclusão, ultramicrotomia e constrastação. In W de Souza, Manual Sobre Técnicas Básicas em Microscopia Eletrônica. Vol. I. Técnicas Básicas. Sociedade Brasileira de Microscopia Eletrônica, Rio de Janeiro, p.44; 2007.

[28] Barreto-Vieira DF, Barth-Schatzmayr OM, Schatzmayr HG. Modelo animal experimental para o estudo da patogênese dos vírus dengue sorotipos 1 e 2, manual de técnicas. Interciência, Rio de Janeiro, 82 p.;2010.

[29] Sesso A. Fixação de sistemas biológicos. In W de Souza, *Manual Sobre Técnicas Básicas em Microscopia Eletrônica*. Vol. I. Técnicas Básicas. Sociedade Brasileira de Microscopia Eletrônica, Rio de Janeiro, p.13-21;2007.

[30] Humprey CD, Pittman EE. A simple methylene blue-azure-II basic fuchsin for epoxy-embedded tissue sections. Stain Technology 1974;49: 9.

[31] Curry A. Electron microscopy and the investigation of a new infectious diseases. Internat J Infect Dis 7: 251-258. Doane FW, Anderson N 1987. Electron Microscopy in Diagnostic Virology, Cambridge University Press, 178p;2003.

Chapter 4

Dechlorination of Chloral Hydrate by *Pseudomonas putida* LF54 which Possesses Biofilm Adhesin Protein LapA

Wanjun Zhang, Huhe, Yuanbai Pan, Yunxiang Cheng and Hiroo Uchiyama

Additional information is available at the end of the chapter

http://dx.doi.org/10.5772/60861

Abstract

Because of the lack of enzymes in critical steps of catabolic pathways, low-molecular-weight halogenated compounds are often recalcitrant to biodegradation. In our previous study, we isolated *Pseudomonas* sp. LF54 (LF54), the first bacterium that has been shown to use chloral hydrate (CH) as sole carbon source by an assimilation pathway in which dechlorination is the critical step. In this study, we identified a transposon (Tn) mutant that can render LF54 defective in CH dechlorination. The molecular characterization of Tn mutants revealed that the transposon insertion sites map to *lapA*. Sequence analyses verified the existence of *lapA* in LF54. Additionally, induced expression of *lapA* in the conditional *lapA* mutant of LF54 further verified that defective *lapA* expression renders LF54 defective in dechlorination. Recent studies have revealed that the largest cell-surface-associated protein LapA, a biofilm adhesin, is able to initiate biofilm formation. This function was also verified in the induced conditional *lapA* mutant and in LF54. Furthermore, we also found out that the defective lapA mutant rendered the variation of bacterial motility. LapA, the largest biofilm adhesin protein of *P. putida*, which influences CH dechlorination and flagella motility, is a novel discovery not previously reported.

Keywords: Chloral hydrate, dechlorination, *lapA*, biofilm, motility

1. Introduction

Chloral hydrate (CH) is synthesized by the chlorination of ethanol. As a sedative and hypnotic drug, CH is used in human and veterinary medicine. Its chemical formula is $C_2H_3Cl_3O_2$. The anhydrous chemical, chloral, is used as an intermediate in the production of insecticides and herbicides such as dichlorvos, naled, methoxychlor, trichlorfon, and trichloroacetic acid [1].

Drinking water is the major exposure route of CH to the general public, as CH is the third by-product formed when drinking water is disinfected with chlorine [2, 3]. Therefore, environments near factories that produce CH and the above-mentioned pesticides might be contaminated. CH is irritating to the skin and mucous membranes. It often causes some symptoms even at the recommended clinical dose [1]. According to the World Health Organization (WHO) guidelines, CH has adverse effect on health and should be limited in drinking water [4]. In addition, CH was reported as a potent genotoxic and carcinogenic compound [3]. Therefore, more efforts should be made to minimize further CH release and contamination in the environment.

Studies of CH were concentrated on animal metabolism, as CH was used as a sedative and hypnotic drug. In animal, the major metabolites are trichloroethanol (TCAol), TCAol glucuronide, trichloroacetic acid (TCA), and a small amount of dichloroacetic acid (DCA) [5-8]. The major excretion pathway of CH metabolite is elimination in the urine. With regard to the bacterial degradation of CH, several cometabolism processes have been reported; methanotrophic-degrading bacteria *Methylocystis* sp. M and *Methylosinus trichosporium* OB3b can transform trichloroethylene into CH, which is further degraded to TCAol and TCA [9, 10]. In our previous study, we successfully isolated *Pseudomonas* sp. LF54 (LF54], the first bacterium reported to date that can use CH as the sole carbon source via an assimilation pathway [11]. This strain transforms CH to TCAol, which is then dechlorinated to 2, 2-dichloroethanol (DCAol), and CO_2 was detected as the end product (Figure 1).

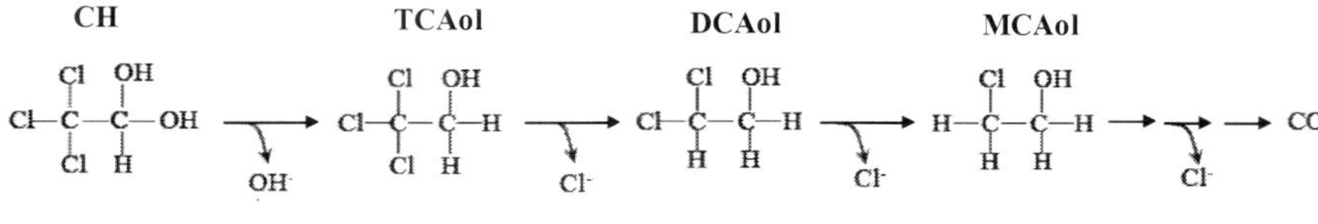

Figure 1. Proposed chloral hydrate biodegradation pathway in *Pseudomonas putida* LF54.

In general, halohydrin dehalogenation is intramolecular substitution that enzymes convert vicinal halohydrins to an epoxide, a proton and a halide ion [12-14]. To detoxify the compounds, halohydrin dehalogenases remove halogen substituents and put them to enter central metabolism. In view of halohydrin dehalogenases properties, the cleavage carbon-halogen bonds started with cloning and expression of three different genes that encode these enzymes [14]. However, the dechlorination in LF54 is reduction that chlorine (Cl) was substituted by hydrogen (H). This is a novel pathway of halohydrin dehalogenation has not been reported to date. Therefore, the further study of LF54 would advantage CH dechlorination and bioremediation. The aim of this Chapter is to identify genes related to CH degradation and to investigate mechanism of CH degradation in *P. putida* LF54 with the method of molecular biology.

CH, chloral hydrate; TCAol, Trichloroethanol; DCAol, Dichloroethanol; MCAol, Monochloroethanol.

2. Identification of genes related to CH degradation

DNA transposition is an important biological phenomenon and also offers versatile tools for genetic analysis of a variety of biological processes. It is applied in creating insertional knockout mutations, providing genetic or physical landmarks for adjacent DNAs, generating the fusions of gene–operon to reporter functions, and locating primer binding sites for the analysis of DNA sequence [15]. Tn5 is one of the simplest and best characterized transposons; there are only Tn5 transposase, transposon DNA flanked by Tn5 inverted repeats, and target DNA are required in vitro transposition [16]. Then electroporation of transposon into a bacterial cell, the reaction proceeds and results in insertion of the transposon DNA at random into the genome to create gene "knockouts" which can block metabolic process.

The transposon used in this chapter was generated with EZ-Tn5™ <KAN-2> Tnp TransposomeTM Kit carrying a kanamycin resistance marker. The transposon can be electroporated into living cells where the transposase is activated by Mg^{2+} in the host's cellular environment, resulting in random insertion of transposon into the host genomic DNA. The transposon-specific primers can be used for bidirectional DNA sequencing from genomic DNA without cloning or locating the transposon insertion sites in target genomic DNA.

In this chapter, in order to identify genes related to CH degradation, a transposon mutant library was generated and Tn mutants (CH degradation-defective) were screened by two screening steps. Transposon insertion sites were sequenced bidirectionally using special sequencing primers from the two sides of the transposon and located using BlastN programs.

2.1. Transposon mutant (Tn mutant) library

LF54 cells were made competent utilizing a microcentrifuge-based procedure [17]. The cells grown in LB mediun were distributed into microcentrifuge and harvested by centrifugation at room temperature for 2 min at 16,000 × g. The cell pellet was washed twice at room temperature and resuspended in 100 μl 300 mM sucrose, which contained average 10^9–10^{10} viable bacteria.

For electroporation, 10 ng of pSUP104 [18] and 20 ng Transposome was mixed with 100 μl electrocompetent cells, and the mixture transferred to 2 mm gap width electroporation cuvette. After applying a pulse (settings: 25 μF; 200 Ω; 2.5 kV on a Bio-Rad GenepulserXcell™, Bio-Rad) [17], the mixture was transferred into 1 mL of room temperature SOC and shaken for 1 h at 30°C; 100 μl of the pSUP104 mixture was plated on Tc plates, and each 100 μl of Transposome mixture was plated on Kana plates. The plates were incubated at 30°C within 24 h.

Controls included cells that were pulsed with added H_2O and were planted on both Tc and Kana plates. Mutagenized cells (Tn mutants) were selected by plating on Kana plate. The strains were stored in 20% glycerin LB media at -80°C.

2.2. Tn mutants screening strategy

Tn mutants (CH dechlorination defective) were screened by two screening steps. The cells were harvested by centrifugation, and washed twice with chloride-free MS medium and were resuspended in a 25 mL serum bottle containing 5 mL of MS medium in proper cells concentration as indicated below. After addition of 1mM CH, the cultures were incubated at 30°C and shaken at 150 rpm for 18 h.

1. **First screening**

The first screening was performed by multistep colorimetric method, as CH degradation is a multistep reaction (Figure 1). The concentration of resuspended cells in this step is 4×10^9 cells mL^{-1}. In weak alkali conditions, 3-methyl-1-phenyl-5-pyrazolone reacts with CH and a brownish red compound is formed and in proportion to CH in 480 nm [19] measured by Varioskan Flash spectral scanning multimode reader (Thermo Fisher Scientific Inc., Waltham, Massachusetts, USA). The chloridion was mixed with a reagent consisting of Hg^{2+}, SCN^-, and Fe^{3+}; a new yellow complex was formed which was measured at 463 nm [20].

2. **Second screening**

The second screening was performed by GC [11]. The concentration of resuspended cells is 2.5×10^{10} cells mL^{-1}. The samples were 10 times concentrated by extraction with t-butyl methyl ether (Wako, Osaka, Japan), and 1,2,3-trichloropropane (Wako) was used as an internal standard. The oven temperature was programmed as follows: 35°C (isothermal) sustained for 7 min then increased to 250°C at a rate of 20°C min^{-1}. The injector and detector temperatures were 270°C.

2.3. Transposon location

Transposon insertion sites were sequenced bidirectionally using sequencing primers, KAN-2 FP-1 and KAN-2 RP-1 (Table 1), specific for the ends of the inserted transposon [21]. The DNA sequencing reactions (two microgram of bacterial genomic DNA and 5 pmoles of primer) were performed using a BigDye™ Terminator Cycle Sequence Kit (Applied Biosystems, Perkin-Elmer, Foster City, CA, USA). The conditions of sequencing reactions are not shown[21]. Sequencing reactions were purified by ethanol precipitation, and resuspended in 10 μl of formamide (Wako, Osaka, Japan). After denaturing at 95°C for 2 min, all products were analyzed with an ABI PRISM® 3130 Genetic Analyzer (Applied Biosystems).

The sequences were assembled and the genomic transposition sites were located using BlastN programs maintained at National Center for Biotechnology Information website (http://blast.ncbi.nlm.nih.gov/).

	Relevant characteristics	source of reference
Stains		
LF54	Wild type *Pseudomonas* sp. LF54	11
PpY101	Wild type *P.putida* PpY101	22
F1	Wild type *P.putida* F1	23
Tn-mus01	LF54 lapA inactivated by transposon, I	This study
Tn-mus03	LF54 lapA inactivated by transposon, I	This study
Tn-mus04	LF54 lapA inactivated by transposon, I	This study
Tn-mus11	LF54 lapA inactivated by transposon, I	This study
Tn-mus13	LF54 lapA inactivated by transposon, I	This study
Tn-mus14	LF54 lapA inactivated by transposon, I	This study
Tn-mus15	LF54 lapA inactivated by transposon, I	This study
Tn control	Tn mutant which can degradated CH as	This study
LF54-lapA	Conditional lapA mutant in LF54 :lapA	This study
E.coli S17-1 λ pir	Used for conjugation of plasmids into L	30
E.coli DH5α	Plamid extraction as control.	Takara
Plamids		
Transposon	EZ-Tn5™ <KAN-2>Tnp Transposom	EPICENTRE Technolog
pSC200	Vector used for single cross-over knock	28,29
pSC200-lapA	lapA fragment inserted into pSC200 fo	This study
pSUP104	the broad-host-range vector Tc^r, Cm^r	
pGEM-T	plamid,Amp^r	
Primers		
EZTN-F	5'- TCTTGCTCGAGGCCGCG-3'	This study
EZTN-R	5'-TTGCATGCC TGCAGGTCG -3'	This study
KAN-2 FP-1 Forward Pri	5'-ACCTACAACAAAGCTCTCATCA	EPICENTRE Technolog
KAN-2 RP-1 Reverse Pri	5'-GCAATGTAACATCAGAGATTTT	EPICENTRE Technolog
lapA-F1	5?- GGAATTCCATATGATGAGCAGC	This study
lapA-R1	5?- GCGGTACCTTATCGAGTACGCC(	This study
lapA-upstream-F	5'-TATCCAGCAGGGGATCGTCA-3'	This study
lapA-R2	5?- TTATCGAGTACGCCCGCAAA-3'	This study
pSC200-F	5'-CGATAGGGCGTCTGCATCC-3'	This study
lapA-R3	5?-TGACGCTACCACTTGTCTGC-3'	This study

Table 1. Bacterial Strains, Plasmids, and Primers Used in the Study

2.4. Results

The transformation efficiency is shown in Table 2. The efficiency of pSUP104 is 1.92×10^{-6}; and the efficiency of transposon is 7.61×10^{-5}. The efficiency of the transposome is almost the same as the plasmid. A library of mutants was generated using Tn mutants in LF54 and stored at -80°C.

Host	DNA	Efficiency (CFU μg^{-1} DNA)
LF54	H_2O (none)	0
	pSUP104 (9.5 kbp)	1.92×10^{6}
	Transposome	7.61×10^{5}

Table 2. Electroporation Efficiency

The first screening for the degradation ability defect was performed by multistep colorimetric method. CH solution (MS medium) as the negative control and LF54 as the positive control were monitored. The results are not shown, as the data are excessive. Disappearance of CH was detected among Tn mutant library; we did not find Tn mutant, in which the process of CH transformed into TCAol was inhibited. According to chloridion release, 96 Tn mutants were picked up (the data are not shown).

The second screening was confirmed using GC; there were 7 Tn mutants that obviously produced less DCAol than LF54 (Figure 2). These strains were designated Tn-mus01, Tn-mus03, Tn-mus04, Tn-mus11, Tn-mus13, Tn-mus14, and Tn-mus15. *P. putida* F1 [22] and PpY101 [23] as the negative control was monitored. All the strains can transform the whole CH into TCAol after 18 h. For strain LF54, approximately 25% TCAol was dechlorinated into DCAol and part of them was converted into CO_2 as the end product [11]. F1 and PpY101 transformed CH into TCAol, but did not subsequently dechlorinate it. However, the DCAol levels in Tn mutants were obviously lower than those in LF54 and closer to F1 and PpY101.

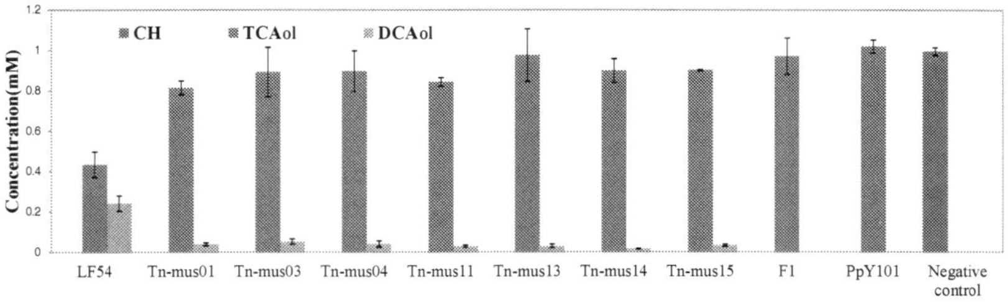

Figure 2. Chloral hydrate (CH) degradation by transposon mutants. The samples were incubated at 30°C and shaken at 150 rpm for 18 h. The initial concentration of CH was 1 mM. CH solution, *P. putida* F1 and PpY101 as the negative control was monitored.

Genomic DNA from 7 Tn mutants that showed inhibited CH dechlorination was purified and directly sequenced. Using primers from each end of the transposon, nearly a kilobase of sequence was assembled from each Tn mutant. All 7 transposon insertion site-flanking sequences (KC686681–KC686687) were mapped to *lapA* in *P. putida* KT2440 [24, 25]; the sequence identities were 91–96% (Table 3).

	Tn-mus01	Tn-mus03	Tn-mus04	Tn-mus11	Tn-mus13	Tn-mus14	Tn-mus15
Aminoacid sites	3157-3471	3392-3717	2929-3242	1663-2003	7334-7651	3388-3658	1252-1573
Sequence identities	92%	94%	94%	94%	91%	95%	94%

Table 3. Sequence Identities of Transposon Mutants to *lapA* in *Pseudomonas putida* KT2440

3. Assay of *lapA* in LF54

Recently, some studies have revealed that the largest cell-surface-associated protein LapA, a biofilm adhesin, is able to initiate biofilm formation and achieve stable, "irreversible" binding to a large variety of surfaces in *P. fluorescens* and *P. putida* [24, 26, 27]. However, no study of *lapA* in connection with the biodegradation of low-molecular-weight chlorinated compounds has been reported to date. The lap genes were conserved among environmental pseudomonads such as P. putida and P. fluorescens, but absent from pathogenic pseudomonads such as P. syringae and P. aeruginosa [24, 27]. In *P. putida* KT2440 (Figure 3), LapA protein is one of the largest bacterial proteins (8,682 amino acids), with an estimated molecular weight of 888 kD. It contains an N-terminal transmembrane region (domain 1), an extensive repetitive region consisting of 9 repeats of 100 amino acids (domain 2) and 29 repeats of 218-225 amino acids (domain 3), and several conserved motifs and domain at the C-terminus of the protein (domain 4).

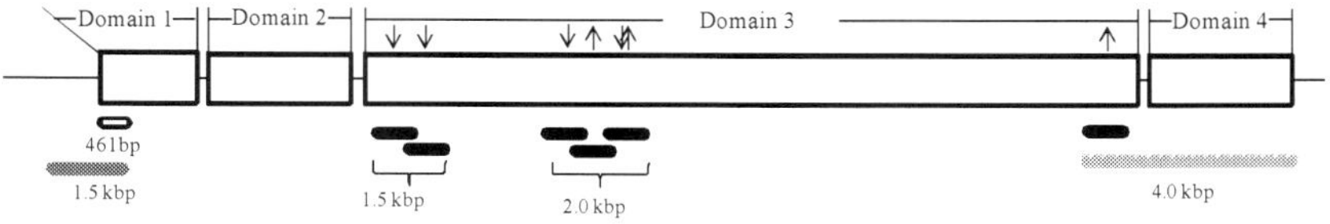

Figure 3. LapA protein in *P. putida* KT2440 and *P. putida* LF54. In *P. putida* KT2440, LapA contains an *N*-terminal transmembrane region (domain 1), an extensive repetitive region consisting of 9 repeats of 100 amino acids (domain 2) and 29 repeats of 218–225 amino acids (domain 3), and several conserved motifs at the *C*-terminus (domain 4). ↓ and ↑ indicate transposon insertion sites and directions. ↓ : Forward direction, ↑ : Backward direction. The bold line indicates *lapA* sequenced in the LF54 chromosome. The dark gray line indicates an upstream fragment including a part of the 5-region of *lapA*; the sequence identity was 89%. The black line indicates the transposon insertion sites flanking the sequence; the sequence identities were 91–96%. The light grey line indicates the 5end sequence of the NODE_189 contig; the sequence identity was 93%. The hollow line indicates the 5-region of *lapA* (461 bp), which was used for constructing the conditional *lapA* mutant.

In this chapter, in order to verify the existence of *lapA* in LF54, we sequence part of *lapA* in LF54. Because the entire *lapA* sequence of LF54 was not acquired from the whole genome sequence analysis, strains defective in *lapA* were assessed to define further the role of *lapA* in CH dechlorination. This strategy created LF54-*lapA*, in which the expression of *lapA* depended on the rhamnose concentration in the medium.

3.1. Existence of *lapA* in LF54

The complete genome sequence of LF54 was performed by Genome Analyzer IIx (Illumina, San Diego, CA) provided by Hokkaido System Science Co., Ltd. The genomic DNA sequence was queried the database using genomic blast programs maintained at National Center for Biotechnology Information website (http://www.ncbi.nlm.nih.gov/sutils/genom_tree.cgi). The upstream fragment includes part of 5′region of *lapA*, amplified by PCR with primers *lapA*-upstream-F, and *lapA*-R2 (Table 1). Primers were synthesized based on the nucleotide sequence from the *Pseudomonas* genome database (http://www.pseudomonas.com). The purified fragment was sequenced using a BigDye™ Terminator Cycle Sequence Kit. The product was analyzed with an ABI PRISM® 3130 Genetic Analyzer (Applied Biosystems). The sequence was used to query the GeneBank database.

3.2. Conditional *lapA* mutant

To construct a conditional *lapA* mutant in LF54 (LF54-*lapA*), the 5-region fragment (461 bp) and the upstream fragment including part of the 5-region of *lapA* amplified by PCR, using the primers *lapA*-F1, *lapA*-R1, *lapA*-upstream-F, and *lapA*-R2 (Table 1) and LF54 genomic DNA as the template were used. Then, the 5-region fragment (461 bp) was digested with *Nde*I/*Kpn*I and ligated into the multiple-cloning site of pSC200 to yield pSC200-*lapA* [28, 29].

The plasmid pSC200-*lapA* was introduced into the mobilizer strain *E. coli* S17-1 λ pir and was then transferred to LF54 by conjugation [30]. Single crossover events were selected on Gm plates to obtain the conditional *lapA* mutant (LF54-*lapA*). The correct insertions were verified by colony PCR with the primers of pSC200-F and *lapA*-R3 (Table 1). DNA fragments were confirmed with electrophoresis on 1.2% agarose gel.

The protein expression of *lapA* was inducted by 0.1% (wt/vol) rhamnose in the medium. CH degradation ability of conditional *lapA* mutant was affirmed by GC method mentioned.

3.3. Results

In this study, the complete genome sequence of LF54 was performed by Genome Analyzer IIx containing 222 contigs. Total contig length is 5,632,841 bp (the data are not shown). The complete genome sequence of LF54 was blasted with 34 *Pseudomonas* genomes databases by Genomic blast program maintained at NCBI website. The almost complete genome sequence of LF54 shows high similarity to *P. putida* strains, the coverage of 86% to *P. putida* F1 (total score = $9.643e^{+06}$), and 85% to *P. putida* KT2440 (total score = $9.132e^{+06}$).

As domains 2 and 3 are an extensive repetitive region, at this stage of the complete genome sequence of LF54 it has not been possible to assemble the complete *lapA* gene. Among the 7

CH dechlorination defective Tn mutants, Tn-mus13 mapped to the end of LapA domain 3, and was also located at 5 end of the NODE_189 contig (294,102 bp, AOUR00000000). Therefore, NODE_189 included a part of domain 3 and the entire domain 4 (Figure 3, light gray line); the sequence identity was 93% to *lapA* in *P. putida* KT2440. In addition, the sequence identity of the upstream fragment including a part of the 5 region of *lapA* (KC686680) was 89% (Figure 3, dark gray line). Combining with sequence flanking the transposon insertions (corresponding to domain 3, Figure 3, black line), the structure of the protein appears to be very similar to its *P. putida* counterpart.

CH degradation ability of LF54-*lapA* was detected by GC analysis (Figure 4). LF54-*lapA* with glucose was used as the negative control, and LF54 was the positive control. LF54-*lapA* can completely transform CH into TCAol after 18 h, irrespective of the presence of rhamnose or glucose. CH dechlorination was inhibited in LF54-*lapA* in the absence of rhamnose, and recovered by rhamnose induction. The sample without rhamnose induction was similar to the glucose control and to the 7 Tn mutants that showed inhibited CH dechlorination.

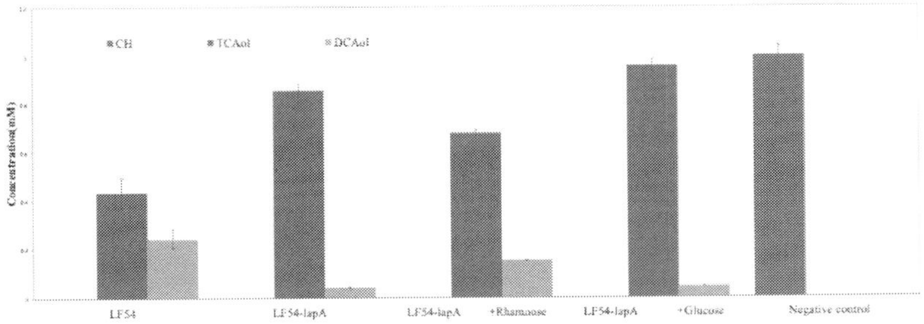

Figure 4. Induction expression of conditional *lapA* mutant (LF54-*lapA*). The samples were incubated at 30°C and shaken at 150 rpm for 18 h. The initial concentration of CH was 1 mM. CH solution (MS medium) as the negative control was monitored.

4. Relationship between CH degradation and biofilm formation

In the past few years, bioremediation mechanisms by *Pseudomonas* for numerous halogenated pollutants have been described and many studies have reported the growth of planktonic cultures under controlled laboratory conditions [31-34]. However, bacteria are mostly found in multicellular communities known as biofilms in the environment, and not in the planktonic state [35, 36]. Biofilm formation in *Pseudomonas* has been proposed to involve a series of regulated steps [37, 26]. Firstly, utilizing flagellar-mediated motility, bacteria swim toward a certain surface and initiate reversible (or transient) attachment. Then, the attached bacteria become irreversibly attached to the surface to form small microcolonies. Finally, the microcolonies develop into a mature biofilm with an architecture separated by fluid-filled channels.

Recent studies have revealed that LapA is able to initiate biofilm formation and achieve stability [24, 26, 27]. However, no study of *lapA* in connection with the biodegradation of CH has been reported to date. We evaluated whether this function is conserved in LF54.

The cultures were started at an optical density of 600 nm (OD_{600}) of 0.01 in glass tubes. After 20 h of growth, the tubes were washed with distilled water and stained with 0.1% crystal violet (Sigma, St. Louis, MO, USA). Following the staining, the tubes were washed twice with distilled water. The tubes were subsequently dried and the crystal violet was dissolved in 96% ethanol for quantification by spectrometry at 595 nm [38].

Biofilm formation in LF54-*lapA* and the Tn-mus13 strain was significantly lower than that in the wild-type strain (shown in Figure 5). The ability to initiate biofilm formation in LF54-*lapA* was restored when *lapA* was induced by rhamnose. These results indicate the relationship between biofilm formation and CH degradation.

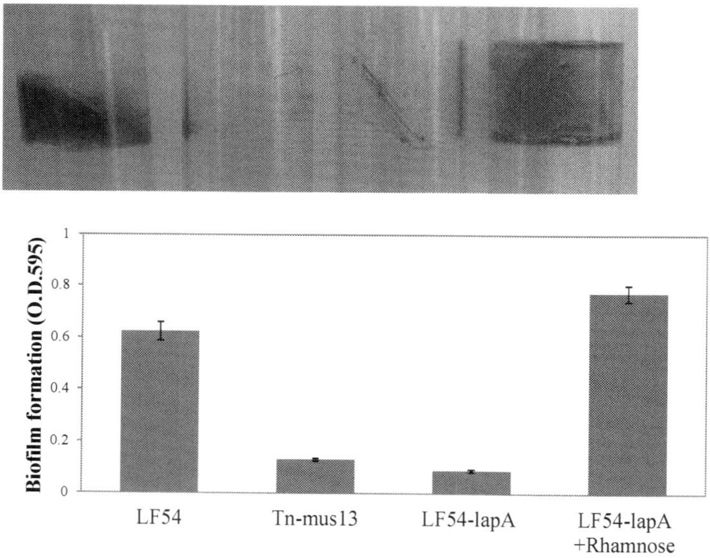

Figure 5. Biofilm formations. The start concentration is 0.01 OD_{600}; the culture was grown in LB medium and shaken at 150 rpm for 20 h.

5. LapA influences bacterial motility

Biofilm research has renewed our interest in bacterial motility [39]. Flagellar and twitching motility are necessary for biofilm development [40]. Biofilm formation of LF54 was verified, and in this chapter, these motilities were observed to verify the influences of LapA.

Twitching motility [41] was measured by stab-inoculating 1.0% agar LB plates with a single bacterial colony. Plates were incubated at 37°C for 2–3days. Flagellar motility [38] including swimming and swarming motility was measured as the diameter of zone travelled by bacteria point-inoculated into 0.3% and 0.5% agar LB plates, respectively. The plates were incubated for 12 h at 30°C. CH was added in a series concentration: 0.1, 0.3, 0.5,1 mM.

Swimming and swarming motility of mutants was faster than LF54 (shown in Figure 6). Although, twitching motility of mutants was also faster than LF54, the difference was not much. Within limits the movement become slowly with the raise of concentration of the CH.

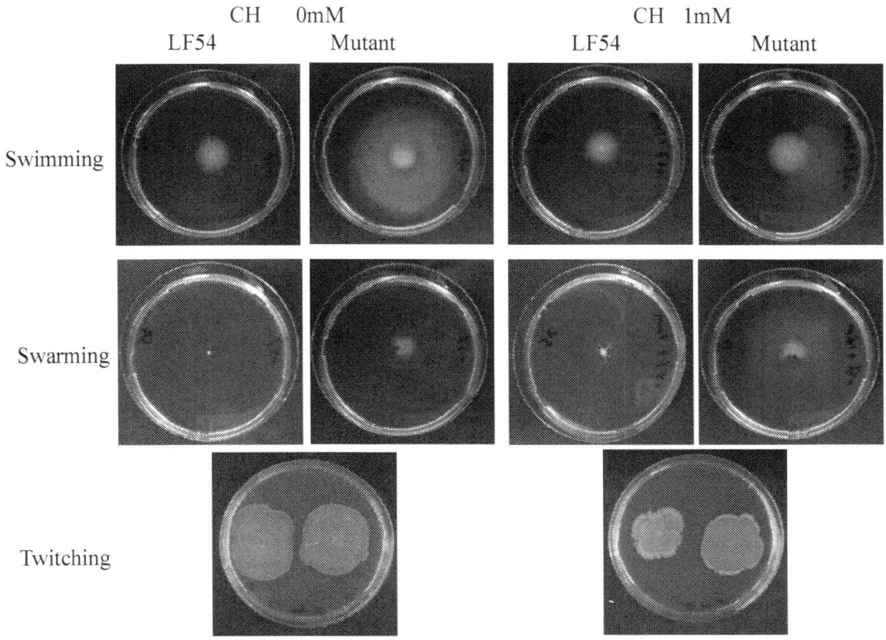

Figure 6. Swimming and swarming motility of mutant was faster than LF54. Twitching motility (pilus movement) of mutant was a little faster than LF54.

Chemotaxis is the movement of organisms toward or away from a chemical. In this study, the defective *lapA* in mutants of LF54 results in the increasing external press to cell. That is, the normal environment for the defective *lapA* mutant, it will amount to over nutrition, the mutant escaped rapidly like the negative chemotaxis. In the CH occasion, the motility appears to like the positive chemotaxis, and the speed was lower than the normal condition. Therefore, we supposed that the other reason of motility change maybe is the biofilm adhesion of mutant changed. The defective adhesion of mutant appears to makes flagellar and twitching motility easily than LF54 and results in faster motility.

6. Conclusion

So, for the first time we demonstrate dechlorination of CH by *Pseudomonas putida* LF54 which possesses Biofilm Adhesin Protein LapA. Although the lapA gene is conserved in *P. putida* strains [9], *P. putida* F1 and PpY101 were not able to dechlorinate CH (Figure 3). Therefore, the LapA protein may have a unique function in LF54; conversely, dechlorination process may involve other factors. The latter possibility is more appealing to us because LapA does not share any similarity with enzymes possessing dehalogenation functions [17]. And we also found out that the defective lapA mutant rendered the variation of bacterial motility. For the future, an interesting challenge will be to verify the interactions between LapA and the CH dechlorination enzyme and to examine whether biofilm formation influences CH dechlorination; and to verify the relationship among the CH dechlorination and LapA and bacterial motility.

Nucleotide sequence accession numbers. The nucleotide sequence data of P. putida LF54 have been submitted to NCBI GenBank under accession numbers KC686680 to KC686687 and AOUR00000000. The version described in this paper is the first version, AOUR01000000.

Acknowledgements

This work was supported by the National Science Foundation for Distinguished Young Scholars of China(31402118), the National Natural Science Foundation of China (31460029), State Key Laboratory of Grassland Agro-ecosystems (SKLGAE201407), the Agricultural Science and Technology Innovation Program (2013GAAS04-04) from Gansu Academy of Agricultural Sciences,a grant-in-aid for scientific research (no. 21310050, to H.U.) and by the CREST program of the Japan Science and Technology Corporation (JST).

We thank Hideaki Nojiri for providing the *P. putida* F1 strain. We thank Leo Eberl for providing the pSC200 plasmid.

Author details

Wanjun Zhang[1], Huhe[2], Yuanbai Pan[3], Yunxiang Cheng[4] and Hiroo Uchiyama[3]

*Address all correspondence to: zhangwanjun0922@yahoo.co.jp

1 College of Environment and Resources, Dalian Nationalities University, Jinzhou New District, Dalian, Liaoning Province, China

2 Institute of Soil and Fertilizer and Save Water Agricultural, Gansu Academy of Agricultural Sciences, Lanzhou, Gansu, China.

3 Graduate School of Life and Environmental Sciences, University of Tsukuba, Tsukuba, Ibaraki, Japan

4 State Key Laboratory of Grassland Agro-ecosystems, College of Pastoral Agriculture Science and Technology, Lanzhou University, Lanzhou, China

References

[1] World Health Organization (WHO). 2000. Chloral hydrate. Concise international chemical assessment document, No25. World Health Organization, Geneva.

[2] Dabrowska A, Nawrocki J. 2009. Controversies about the occurrence of chloral hydrate in drinking water. *Water Res*. 43:2201-2208.

[3] Meric S. 2007. Toxicity of disinfectants and disinfection by-products: what did we learn to explore more? p 231-249. *In* A. Nikolaou, L. Rizzo, H. Selcuk (eds.), *Control of Disinfection By-Products in Drinking Water Systems*. Nova Science Publishers, Inc., New York, NY.

[4] World Health Organization (WHO). 2004. *Guidelines for Drinking Water Quality*. World Health Organization, Geneva. Volume 1 – Recommendations, third edition.

[5] Beland FA, Schmitt TC, Fullerton NF, Young JF. 1998. Metabolism of chloral hydrate in mice and rats after single and multiple doses. *J Toxicol Environ Health A*. 54:209-226.

[6] Allen BC, Fisher JW. 1993. Pharmacokinetic modeling of trichloroethylene and trichloroacetic acid in humans. *Risk Anal*. 13:71-86.

[7] Ketcha MM, Stevens DK, Warren DA. Bishop CT. Brashear WT. 1996. Conversion of trichloroacetic acid to dichloroacetic acid in biological samples. *J Anal Toxicol*. 20:236-241.

[8] Abbas R, Seckel CS, Kidney JK, Fisher JW. 1996. Pharmacokinetic analysis of chloral hydrate and its metabolism in B6C3F1 mice. *Drug Metab Dispos*. 24:1340-1346.

[9] Nakajima T, Uchiyama H, Yagi O, Nakahara T. 1992. Novel metabolite of trichloroethylene in a methanotrophic bacterium, *Methylocystis sp*. M, and hypothetical degradation pathway. *Biosci Biotech Biochem*. 56:486-489.

[10] Oldenhuis R, Vink RL, Janssen DB, Witholt B. 1989. Degradation of chlorinated aliphatic hydrocarbons by *Methylosinus trichosporium* OB3b expressing soluble methane monooxygenase. *Appl Environ Microbiol*. 55:2819-2826.

[11] Huhe, Nomura N, Nakajima T, Uchiyama H. 2011. Assimilative and co-metabolic degradation of chloral hydrate by bacteria and their bioremediation potential. *J Biosci Bioeng*. 111:448-453.

[12] Lutje Spelberg JH, van Hylckama Vlieg JET, Bosma T, Kellogg RM, Janssen DB. 1999. A tandem enzyme reaction to produce optically active halohydrins, epoxides and diols. *Tetr Assymm*. 10:2863-2870.

[13] de Jong RM, Tiesinga JJ, Rozeboom HJ, Kalk KH, Tang L, Janssen DB, Dijkstra BW. 2003. Structure and mechanism of a bacterial haloalcohol dehalogenase: a new variation of the short-chain dehydrogenase/reductase fold without an NAD(P)H binding site. *EMBO J*. 22:4933-4944.

[14] van Hylckama Vlieg JET, Tang L, Lutje Spelberg JH, Smilda T, Poelarends GJ, Bosma T, van Merode AE, Fraaije MW, Janssen DB. 2001. Halohydrin dehalogenases are structurally and mechanistically related to short-chain dehydrogenases/reductases. *J Bacteriol*. 183:5058-5066.

[15] Goryshin IY, Jendrisak J, Hoffman LM, Meis R, Reznikoff WS. 2000. Insertional transposon mutagenesis by electroporation of released Tn5 transposition complexes. *Nat Biotechnol*. 18:97-100.

[16] Fernandes PJ, Powell JA, Archer JA. 2001. Construction of *Rhodococcus* random mutagenesis libraries using Tn5 transposition complexes. *Microbiology*. 147:2529-2536

[17] Choi KH, Kumar A, Schweizer HP. 2006. A 10-min method for preparation of highly electrocompetent *Pseudomonas aeruginosa* cells: application for DNA fragment transfer between chromosomes and plasmid transformation. *J Microbiol Methods*. 64:391-397.

[18] Kurabayashi T, Iwasaki K, Uchiyama H, Nakamura K, Tanaka H, Yagi O. 1997. Characteristics of *Escherichia coli* HB101 and *Pseudomonas putida* PpY101 harboring a recombinant plasmid with tandem insertion of the mercury resistance operon. *Biosci Biotechnol Biochem*. 61:1187-1189.

[19] Water quality. 2000. Determination of trichloro-aldehyde.3-methyl-1-phenyl-5-pyrazolone spectrophotometric method. HJ/T 50-1999. http://english.mep.gov.cn/standards_reports/standards/water_environment/method_standard2/.

[20] Bergmann JG, Sanik J. 1957. Determination of trace amounts of chlorine in naphtha. *Anal Chem*. 29:241-243.

[21] Hoffman L, Jendrisak J. 1999. Use of EZ::TN™ Transposomes™ for genetic analysis and direct sequencing of bacterial genomic DNA. EPICENTRE Technologies. http://www.arb-ls.com/products/epicentre_forum/6_3.pdf.

[22] Spain JC, Gibson DT. 1988. Oxidation of substituted phenols by *Pseudomonas putida* F1 and *Pseudomonas* sp. strain JS6. *Appl Environ Microbiol*. 54:1399-1404.

[23] Kurabayashi T, Iwasaki K, Uchiyama H, Nakamura K, Tanaka H, Yagi O. 1997. Characteristics of *Escherichia coli* HB101 and *Pseudomonas putida* PpY101 harboring a recombinant plasmid with tandem insertion of the mercury resistance operon. *Biosci Biotechnol Biochem*. 61:1187-1189.

[24] Hinsa SM, Espinosa-Urgel M, Ramos JL, O'Toole GA. 2003. Transition from reversible to irreversible attachment during biofilm formation by *Pseudomonas fluorescens* WCS365 requires an ABC transporter and a large secreted protein. *Mol Microbiol*. 49:905-918.

[25] Nelson KE, Weinel C, Paulsen IT, Dodson RJ, Hilbert H, Martins dos Santos VA, Fouts DE, Gill SR, Pop M, Holmes M, Brinkac L, Beanan M, DeBoy RT, Daugherty S, Kolonay J, Madupu R, Nelson W, White O, Peterson J, Khouri H, Hance I, Chris Lee P, Holtzapple E, Scanlan D, Tran K, Moazzez A, Utterback T, Rizzo M, Lee K, Kosack D, Moestl D, Wedler H, Lauber J, Stjepandic D, Hoheisel J, Straetz M, Heim S, Kiewitz C, Eisen JA, Timmis KN, Düsterhöft A, Tümmler B, Fraser CM. 2002. Complete genome sequence and comparative analysis of the metabolically versatile *Pseudomonas putida* KT2440. *Environ Microbiol*. 4:799-808.

[26] Gjermansen M, Nilsson M, Yang L, Tolker-Nielsen T. 2010. Characterization of starvation-induced dispersion in *Pseudomonas putida* biofilms: genetic elements and molecular mechanisms. *Mol Microbiol*. 75:815-826.

[27] Ivanov IE, Boyd CD, Newell PD, Schwartz ME, Turnbull L, Johnson MS, Whitchurch CB, O'Toole GA, Camesano TA. 2012. Atomic force and super-resolution microscopy support a role for LapA as a cell-surface biofilm adhesin of *Pseudomonas fluorescens*. *Res Microbiol*. 163:685-691.

[28] Ortega XP, Cardona ST, Brown AR, Loutet SA, Flannagan RS, Campopiano DJ, Govan JR, Valvano MA. 2007. A putative gene cluster for aminoarabinose biosynthesis is essential for *Burkholderia cenocepacia* viability. *J Bacteriol*. 189:3639-3644.

[29] Werneburg M, Zerbe K, Juhas M, Bigler L, Stalder U, Kaech A, Ziegler U, Obrecht D, Eberl L, Robinson JA. 2012. Inhibition of lipopolysaccharide transport to the outer membrane in *Pseudomonas aeruginosa* by peptidomimetic antibiotics. *Chembiochem*. 13:1767-1775.

[30] de Lorenzo V, Timmis KN. 1994. Analysis and construction of stable phenotypes in gram-negative bacteria with Tn5- and Tn10-derived minitransposons. *Methods Enzymol*. 235:386-405.

[31] Arif MI, Samin G, van Leeuwen JG, Oppentocht J, Janssen DB. 2012. Novel dehalogenase mechanism for 2,3-dichloro-1-propanol utilization in *Pseudomonas putida* strain MC4. *Appl Environ Microbiol*. 78:6128-6136.

[32] Goswami M, Recio E, Campoy S, Martín JF, Coque JJ. 2007. Environmental significance of *O*-demethylation of chloroanisoles by soil bacterial isolates as a mechanism

that improves the overall biodegradation of chlorophenols. *Environ Microbiol.* 9:2512-2521.

[33] Travkin VM, Solyanikova IP, Rietjens IM, Vervoort J, van Berkel WJ, Golovleva LA. 2003. Degradation of 3,4-dichloro- and 3,4-difluoroaniline by *Pseudomonas fluorescens* 26-K. *J Environ Sci Health B.* 38:121-132.

[34] Garbi C, Casasús L, Martinez-Alvarez R, Ignacio Robla J, Martín M. 2006. Biodegradation of oxadiazon by a soil isolated *Pseudomonas fluorescens* strain CG5: Implementation in an herbicide removal reactor and modelling. *Water Res.* 40:1217-1223.

[35] Costerton JW, Lewandowski Z, Caldwell DE, Korber DR, Lappin-Scott HM. 1995. Microbial Biofilms. *Ann Rev Microbiol.* 49:711–745.

[36] Davey ME, O'Toole GA. 2000. Microbial biofilms: from ecology to molecular genetics. Microbiol. *Mol Biol Rev.* 64:847-867.

[37] O'Toole G, Kaplan HB. 2000. Kolter R. Biofilm formation as microbial development. *Ann Rev Microbiol.* 54:49-79.

[38] O'Toole GA, Kolter R. 1998. Initiation of biofilm formation in *Pseudomonas fluorescens* WCS365 proceeds via multiple, convergent signalling pathways: a genetic analysis. *Mol Microbiol.* 28:449-461.

[39] Daniel B. Kearns. 2010. A field guide to bacterial swarming motility. *Nat Rev Microbiol.* 8:634–644.

[40] O'Toole GA, Kolter R. 1998. Flagellar and twitching motility are necessary for *Pseudomonas aeruginosa* biofilm development. *Mol Microbiol.* 30:295-304.

[41] Whitchurch CB, Hobbs M, Livingston SP, Krishnapillai V, Mattick JS. 1991. Characterisation of a *Pseudomonas aeruginosa* twitching motility gene and evidence for a specialised protein export system widespread in eubacteria. *Gene.* 101:33-44.

Chapter 5

An Insight Into Beneficial *Pseudomonas bacteria*

Galina Novik, Victoria Savich and Elena Kiseleva

Additional information is available at the end of the chapter

http://dx.doi.org/10.5772/60502

Abstract

Pseudomonas is a widespread bacterial genus embracing a vast number of species. Various genosystematic methods are used to identify *Pseudomonas* and differentiate these bacteria from species of the same genus and species of other genera. Ability to degrade and produce a whole spectrum of compounds makes these species perspective in industrial applications. It also makes possible to use various media, including wastes, for cultivation of *Pseudomonas*. Pseudomonads may be applied in bioremediation, production of polymers and low-molecular-weight compounds, biocontrol. Recent studies open up new frontiers for further use of *Pseudomonas* in various areas.

Keywords: *Pseudomonas* bacteria, physiology, taxonomy, application

1. Introduction

Pseudomonas is one of the most studied species of bacteria. They were first identified at the end of 19th century by Migula as Gram-negative, rod-shaped and polar-flagellated bacteria. Since that time description of genus *Pseudomonas* has widened; development of new methods allowed to study in detail the morphology and physiology of these bacteria. However, the morphological characteristics of *Pseudomonas* are common to many bacterial genera and so are of little value in the positive identification of members of the genus. Advanced nucleic acid-based methods allow to differentiate it from other similar genera and reveal taxonomic relationships among various bacterial species including *Pseudomonas*.

Genus *Pseudomonas* is represented by species that occupy a wide range of niches owing to metabolic and physiological diversity. This diversity allows pseudomonads to adapt to challenging environment, resist to adverse conditions caused by abiotic and biotic factors such as high and low temperature, moisture, oxygen and nutrients availability, antibiotics, etc. Elevated resistance provides for ubiquitous distribution of *Pseudomonas* in soil and water, as well as plant growth-promoting rhizobacteria (PGPR), animal and plant pathogens. The bacterium is capable of utilizing a broad spectrum of organic compounds as sources of carbon and energy, hence it is able to colonize habitats where nutrients are limited.

Diversity of *Pseudomonas* determines vast research interest in this genus. Some species like *P. aeruginosa* are opportunistic human pathogens showing enhanced antibiotic resistance, so that studies of pathogenic strains are centered on mechanisms of this antibiotic resistance. Other species are able to degrade a number of compounds that are toxic or recalcitrant to other bacterial species, or produce a wide range of secondary metabolites and biopolymers. It makes these strains perspective for industrial applications.

2. Morphology and physiology of *Pseudomonas* bacteria

Pseudomonas are Gram-negative, aerobic, motile by one or several polar flagella, non-spore-forming straight or slightly curved rods. In addition to the polar flagella, some species (*P. stutzeri, P. mendocina*) have shorter lateral flagella. Solid media favor the formation of lateral flagella which are closely related with swarming of cells on solid surfaces [1]. The number of flagella has taxonomic importance. Most *P. aeruginosa* cells carry only one flagellum, although some cells hold two or three flagella. *P. alcaligenes, P. mendocina, P. pseudoalcaligenes,* and *P. stutzeri* are also characterized by a single flagellum. The majority of species possess more than one flagella [2].

Some *Pseudomonas* species also form pili (*P. aeruginosa, P. alcaligenes, P. syringae*). Type IV pili of *P. aeruginosa* similar to pili of other pathogenic bacteria are involved in cell adhesion to epithelial cells [3, 4]. Pili are essential for the normal development of *P. aeruginosa* biofilms, and they also function as receptors for bacteriophage binding [5-7]. The adhesive region is located at the tip of the pilus. Pili of phytopathogenic *P. syringae* serve as a conduit for the long-distance translocation of effector proteins in plant cells [8].

Bacterial cells don't produce prosthecae and aren't surrounded by sheaths, but they can form biofilms that provide attachment of cells to the substrate and increase stability under adverse conditions [9].

Another important *Pseudomonas* feature is production of variety of pigments. Character of pigmentation remains significant factor among the diagnostic traits of *Pseudomonas*. Pigments may be soluble in water and diffusible into the medium or may be associated with the cells. Pseudomonads can produce diffusible pigments that fluoresce in short wavelength (254 nm) ultraviolet light. Some of these pigments, like yellow-green pyoverdine (fluorescein), are siderophores that play an important physiological role in satisfying the iron requirement. The synthesis of pyoverdine is strongly related to iron starvation. It can be demonstrated by cultivating the bacteria in media such as King's medium B. Pyoverdine binds iron (III) ions very tightly, and that ferripyoverdine complex is actively transported into the bacterial cell [10, 11]. Pyoverdine from *P. aeruginosa* is essential for virulence in animal models [12]. Pyoverdine also can be a tool for identification of *Pseudomonas* because each genomic group is characterized by a specific pyoverdine [13]. Other pigments produced by species of *Pseudomonas* include pyocyanin (*P. aeruginosa*, blue color), pyorubin (*P. aeruginosa*, red color), chlororaphin (*P. chlororaphis*, green color), pyomelanin (*P. aeruginosa*, brown/black color). *P. mendocina* is able to produce carotenoid pigment [14, 15].

Pseudomonas are aerobic bacteria, but in some cases they can use nitrate as alternate electron acceptor and carry out denitrification (*P. aeruginosa*, *P. stutzeri*, and some *P. fluorescens* biovars), reducing nitrate to N_2O or N_2. Additionally, *P. chloritidismutans* can utilize chlorate (ClO_3^-) as an alternative energy-yielding electron acceptor [16].

Pseudomonas tends to utilize organic acids in preference to more complex organic compounds. It represses many inducible peripheral catabolic enzymes. Most of *Pseudomonas* species have incomplete glycolytic pathways, lacking 6-phosphofructokinase, therefore sugars and organic acids are dissimilated prevalently via the Entner-Doudoroff pathway. Representatives of the genus can utilize common monosaccharides (glucose, fructose, galactose, l-arabinose), but growth of some species (*P. stutzeri, P. mendocina, P. syringae*) may be slow. Most hexoses and related compounds are also degraded by the Entner-Doudoroff pathway and various peripheral pathways [2, 14, 17].

Strains of *Pseudomonas* can grow in minimal media with ammonium ions or nitrate as nitrogen source and a single organic compound as the sole carbon and energy source, not requiring organic growth factors. Some species like strains of phytopathogenic *P. syringae* grow very slowly in comparison with strains of the saprophytic species, but their growth can be enhanced by addition of small amounts of complex organic materials (yeast extract, peptones). Significant systematic feature of *Pseudomonas* is inability to accumulate polyhydroxybutyrate, but polyhydroxyalkanoates of monomer lengths higher than C_4 may be accumulated when growing on alkanes or gluconate. Optimal temperature for growth is approximately 28ºC, although some species can grow at 4ºC or 41ºC. Most species can't tolerate acid conditions (pH 4.5 or lower) [14].

Members of the genus *Pseudomonas* are known for their degrading ability on a whole range of substrates, like hydrocarbons, aromatic compounds, and their derivatives. Some of these compounds are natural (toluene, styrene, naphthalene, phenol), other compounds are final products or intermediates from industrial activities (polychlorobiphenyls, dioxins, nitrotoluenes). A considerable number of these compounds is toxic to microorganisms of other groups and to higher organisms. Research revealed 11 central pathways to which many different peripheral pathways converge. Peripheral pathways transform substrates into a few central intermediates (usually dihydroxybenzenes or dihydroxyaromatic acids), which are then ring-cleaved and converted to tricarboxylic acid (TCA) cycle intermediates through the corresponding central pathways. *P. putida* contains 9 out of the 11 identified central pathways, which is in agreement with the wide range of niches that this species can colonize. The ability to degrade aromatic compounds is a strain-specific feature, therefore several pathways that are found in some strains are missing in other strains of the same species [18].

The β-ketoadipate pathway is the most widespread *Pseudomonas* pathway of the degradation of aromatic compounds. It includes *ortho*-cleavage protocatechuate (*pca* genes) and catechol (*cat* genes) branches. Both *cat* and *pca* branches are usually present in most organisms, but the *cat* branch is absent in the three available genomes of *P. syringae*. Quinate, *p*-hydroxybenzoate, and phenylpropenoids (p-coumarate, caffeate, cinnamate, ferulate, etc.) are degraded via the intermediate protocatechuate. Protocatechuate is cleaved by 3,4-dioxygenase to carboxy-*cis,cis*-muconate that is converted to β-ketoadipate enol-lactone by PcaC and PcaD enzymes. The *pca* genes are gathered in a single cluster in *P. fluorescens*, but they are organized in different clusters in other *Pseudomonas* strains [19]. Benzoate, tryptophan, aniline, salicylate, naphthalene, biphenyl, phenol, benzene, toluene, 4-nitrotoluene, and nitrobenzene are degraded via catechol. Benzoate is an intermediate in the catabolism of several aromatic compounds. Its degradation to catechol involves the *benABCD* genes which were identified in the *Pseudomonas* genomes carrying *cat* genes. *Cat* genes encode enzymes involved in catechol *ortho*-cleavage. *CatA*, *catB*, and *catC* encode catechol 1,2-dioxygenase, *cis,cis*–muconate cycloisomerase, and muconolactone isomerase, respectively. The *ben* and *cat* genes are located together in the genomes of *P. fluorescens*, *P. aeruginosa*, *P. stutzeri*, and *P. entomophila*. Reverse situation is observed in most *P. putida* strains.

Genus *Pseudomonas* also displays other metabolic pathways for aromatic compounds: phenylacetyl-CoA (phenylethylamine, phenylethanol, styrene, tropate), homogentisate (phenylalanine, tyrosine), gentisate (salicylate, 3-hydroxybenzoate, *m*-cresol), homoprotocatechuate (4-hydroxyphenylacetate), nicotinate (nicotinic acid), etc [18].

3. Taxonomy and identification of *Pseudomonas* bacteria

The genus *Pseudomonas* belongs to phylum Proteobacteria, class Gammaproteobacteria, order Pseudomonadales, family Pseudomonadaceae with type species *P. aeruginosa*. At present the genus includes about 216 species with 18 subspecies and the number of species constantly increases [20].

The identification of *Pseudomonas* is a necessary step preceding further use of these bacteria because of pathogenicity of some strains to plants and animals, including humans, and its wide metabolic diversity. Since the discovery of *Pseudomonas*, a large number of species was assigned to the genus. First classification of *Pseudomonas* species was based on phenotypic characteristics. The most significant work providing phenotypic description of this genus was performed by Stanier et al. Strains of different species were subjected to many phenotypic tests, the most important of which was the nutritional screening [21].

In the 1960s studies on nucleic acid similarity have been started. DNA–DNA hybridization (DDH) has shown high degree of genomic heterogeneity among the species assigned to the genus [14, 22]. DDH is a universal technique that could offer truly genome-wide comparisons between organisms, but it demands large quantities of high-quality DNA (in comparison with PCR-based techniques). It makes DDH time-consuming and labour-intensive [23].

Evidence of the high level of conservatism among ribosomal RNA molecules [24, 25] allowed to divide this genus into five rRNA groups using rRNA–DNA hybridization [26]. Only rRNA group I that included the type species *P. aeruginosa,* all the fluorescent (*P. fluorescens, P. putida, P. syringae*), and some non-fluorescent species (*P. stutzeri, P. alcaligenes, P. pseudoalcaligenes, P. mendocina*) reserved the name *Pseudomonas*. Later the residuary rRNA groups were affiliated to other genera. The species of rRNA group II were transferred to the genera *Burkholderia* and *Ralstonia,* the species of rRNA group III were transferred to the genera *Acidovorax, Comamonas,* and *Hydrogenophaga,* the species of rRNA group IV and group V were transferred to the genera *Brevundimonas* and *Stenotrophomonas,* respectively [27-33].

Sequential development of molecular methods has emphasized the role of 16S rRNA in the identification and classification of bacteria, including *Pseudomonas*. Reasons that allow wide use of 16S rRNA for taxonomic studies include: presence in almost all bacteria, often existing as a multigene family, or operons; the function of the 16S rRNA gene over time has not changed, suggesting that random sequence changes are a more accurate measure of time (evolution); the 16S rRNA gene (1500 bp) is large enough for informatics purposes [34]. 23S rRNA is excessively conserved and 5S rRNA is too small for research.

As a result of 16S rRNA sequencing by Moore et al., the genus *Pseudomonas* was grouped into 2 distinct intrageneric divisions. These divisions are designated the "*P. aeruginosa* intrageneric cluster" and the "*P. fluorescens* intrageneric cluster" including four (*P. aeruginosa, P. resinovorans, P. mendocina, P. flavescens*) and five (*P. fluorescens, P. syringae, P. cichorii, P. putida, P. agarici*) lineages, respectively. Sequence similarities between the species ranged from 93.3% (between *P. cichorii* and *P. cirtonellolis*) to 99.9% (between *P. oloevorans* and *P. pseudoalcaligenes*). It was observed that 148 positions of 16S rRNA were variable among 1492 nucleotide positions, and 65 positions of these nucleotides were located within three hypervariable regions. Approximately 44% of the total gene sequence variability of *Pseudomonas* species occurs in 6% of the 16S rRNA sequence. Regions other than the variable regions are crucial for ribosome functions [35]. In the research of Anzai et al. genus *Pseudomonas* was divided into two clusters using 16S rRNA sequencing. Six groups were defined within the first cluster: *P. syringae, P. chlororaphis, P. fluorescens, P. stutzeri, P. aeruginosa,* and *P. putida* groups. *P. agarici* and *P. asplenii* belong to first cluster, but they were not included into any group. The second cluster contained only *P. pertucinogena* group [36].

Although 16S rRNA gene sequencing is useful for classification and identification, it has some resolution problems at the genus and species level. These problematic groups include the family *Enterobacteriaceae* (in particular, *Enterobacter* and *Pantoea*), the *Acinetobacter baumannii-A. calcoaceticus* complex, genera *Achromobacter, Stenotrophomonas, Actinomyces,* and some species such as *Bacillus anthracis, Bacillus cereus, Bacillus globisporus, Bacillus psychrophilus, Burkholderia cocovenenans, Burkholderia gladioli, Burkholderia pseudomallei, Burkholderia thailandensis, Neisseria cinerea, Neisseria meningitides, Pseudomonas fluorescens, Pseudomonas jessenii, Streptococcus mitis, Streptococcus oralis, Streptococcus pneumonia,* etc. Some of these problems are related to bacterial nomenclature and taxonomy; others are related to sequence identity and very high similarity scores. Some species, like *Aeromonas veronii,* could contain up to six copies of the 16S rRNA gene that differ by up to 1.5% among themselves which might cause identification problems [37].

Some conservative genes such as *gyrB* (DNA gyrase B subunit) and *rpoD* (σ^{70} factor) also can be used for the identification because protein-encoding genes evolve much faster than rRNA genes and provide higher resolution of intrageneric relationships than 16S rRNA sequencing. Using these genes, two major intrageneric clusters were identified. These intrageneric divisions are consistent with the clusters that have been defined using 16S rRNA gene sequence analysis by Moore, but phylogenetic relationships within the clusters differ in comparison with 16S rRNA sequencing. *GyrB* and *rpoD* nucleotide sequences can be also used to design specific PCR primers due to the high evolution rates [38, 39]. *OprI, rpoB, atpD, carA, recA,* and *oprF* also can serve as alternative phylogenetic markers when studying *Pseudomonas* taxonomy [40-43].

Another recently introduced method for taxonomic investigations of bacteria is multilocus sequence typing/analysis (MLST/MLSA). MLSA is a molecular typing method that consists of sequencing 400-600 bp long fragments of some housekeeping genes, i.e., genes that are present in most bacteria. MLSA has two important advantages over 16S rRNA sequencing: 1) the higher variability of housekeeping genes as compared to the 16S rRNA sequence and increased length of the total analyzed sequence even allow differentiation of strains; 2) sequencing of some genes reduces the risk that horizontal gene transfer obscures the resulting phylogeny [44]. According to the recent MLSA research (16S rRNA, *gyrB, rpoB,* and *rpoD* genes) the genus *Pseudomonas,* as before, is divided into two lineages (*P. aeruginosa* and *P. fluorescens*), which are subdivided into three and nine groups, respectively. The *P. oryzihabitans* group (two species) and the type strains of *P. luteola, P. pachastrellae,* and *P. pertucinogena* are the most phylogenetically distant from all other *Pseudomonas* and therefore they aren't included in these lineages [45].

In addition to sequencing of different genes it's possible to use a number of other methods. Restriction fragment length polymorphism (RFLP) is related to the polymorphic nature of the locations of restriction enzyme sites within defined genetic regions. As a result of RFLP, restriction profile is revealed. RFLP procedure is simple in manipulation and it doesn't require sequence information allowing to identify bacteria at species or subspecies level. On the other hand, it's time consuming and requires large amounts of DNA. The method was applied to determine genomovars and biotypes of various *Pseudomonas* species using 16S rRNA or 16S-23S spacer regions [46, 47]. The intergenic 16S-23S internally transcribed spacer (ITS1) regions are less susceptible to selection pressure because of their non-coding function and should have accumulated a higher percentage of mutations than the rRNA genes [46].

It's possible to use polymerase chain reaction-reverse cross-blot hybridization (PCR-RCBH) in detection and identification studies. 16S-23S intergenic spacer region was amplified and used in hybridization assay with specific oligonucleotide probes to fluorescent pseudomonads and certain species of the genus. Positive reactions were observed if studied bacteria at least belonged to genus *Pseudomonas*. It was demonstrated that the identification of pseudomonads by PCR-RCBH is highly specific and less time-consuming than the conventional bacterial culture method [48].

Pulsed-field gel electrophoresis (PFGE) can be used for differentiation and identification of single strains [49, 50]. PFGE is often considered the "gold standard" of molecular typing methods. PFGE has the high discriminatory power, however, this method is time-consuming and labour-intensive, and some point mutations can change banding patterns, resulting in misleading results [51]. Enterobacterial repetitive intergenic consensus PCR (ERIC-PCR) is also an effective method for identification of *Pseudomonas* genotypes. ERIC-PCR is quick, easy to perform and cost effective, but it has low reproducibility compared to PFGE [52-54].

As mentioned above, some chemotaxonomic markers like pyoverdines also can be taxonomic tools for the identification of *Pseudomonas*. Strains belonging to a well-defined genomic group produce an identical pyoverdine, and each genomic group is characterized by a specific pyoverdine. The same conclusions are valid for nonfluorescent *Pseudomonas* species and their siderophores. Strains are analyzed by two siderotyping methods: siderophore isoelectrofocusing and siderophore-mediated iron uptake. Correlation between DNA-DNA hybridization and siderotyping data was established. Compared to conventional phenotypic and genomic methods, siderotyping is the fast, accurate, and easy-to-perform technique allowing to identify at the species level. Two siderotyping methods can be improved by mass spectrometric determination of the molecular mass of pyoverdines [13, 55, 56].

Another possible tool for *Pseudomonas* taxonomy is fluorescence spectroscopy. In the study of Tourkya et al. analysis of emission spectra of three intrinsic fluorophores (NADH, tryptophan, and the complex of aromatic amino acids and nucleic acid) allowed to clearly discriminate *Pseudomonas* at genus level from *Burkholderia*, *Xanthomonas* and *Stenotrophomonas*. These results correlate with the classification based on 16S rRNA comparison. Fluorescence spectroscopy also allowed to discriminate *P. lundensis*, *P. taetrolens*, *P. fragi*, *P. chlororaphis*, and *P. stutzeri* species from the others. Clustering of these species is also concordant with data from 16S rRNA gene sequence comparison affiliating the four species to the same *P. chlororaphis* group [57].

There are many methods allowing to identify and classify the *Pseudomonas* genus, but gene sequencing procedures proved the most advanced and sophisticated. Great diversity of genus *Pseudomonas* urges further progress of taxonomic methodology.

4. Criteria for selection of *Pseudomonas* bacteria

As mentioned above, pseudomonads are able to degrade a broad spectrum of compounds. They are also characterized by an enormous biosynthesis capacity resulting in the production of a wide range of secondary metabolites and biopolymers. Ability to degrade and synthesize various substances is a vital technological merit of *Pseudomonas*. It promotes practical interest in various biotechnological processes such as bioremediation, production of polymers, biotransformation, synthesis of low-molecular-weight compounds and recombinant proteins, biocontrol agents [58]. The above-mentioned applications demand formulation of criteria for selection of pseudomonads.

5. Safety of *Pseudomonas* bacteria

Pseudomonas is known to display a range of pathogenic and toxicological characteristics in regard to humans, animals, and plants. The infections pseudomonads cause to humans are generally opportunistic. Individuals most at risk from *Pseudomonas* infection are the immunocompromised, patients with cystic fibrosis, and patients suffering major trauma or burns. The predominant *Pseudomonas* species isolated from clinical sources are *P. aeruginosa* [59]. *P. aeruginosa* is an opportunistic pathogen that may induce severe infections in humans and other vertebrates. Some *P. aeruginosa* strains, like PA14, also cause disease in a variety of nonvertebrate hosts, including plants, *Caenorhabditis elegans*, and the greater wax moth, *Galleria mellonella* [60]. The other *Pseudomonas* infection cases are rare.

Important feature of *Pseudomonas* is antibiotic resistance. Antibiotic resistance in the bacterial community constantly increases, and more multiple drug resistant strains appear. The best studied organism among pseudomonads is *P. aeruginosa*. The species is known for multiple drug resistance. *P. aeruginosa* has acquired resistance via multiple mechanisms, including production of β-lactamases and carbapenemases, upregulation of multidrug efflux pumps, and cell wall mutations leading to a reduction in porin channels [61].

Antibiotics used to treat *P. aeruginosa* infections have to cross the cell wall to reach their targets. The resistance of *P. aeruginosa* to these antibiotics is connected, first of all, with low permeability of the outer cell membrane and the efficient removal of antibiotics by efflux pumps. The above-mentioned mechanisms are common components of the resistance phenotype for β-lactams, aminoglycosides, and quinolone antibiotics. The agents that break down the outer-membrane permeability barrier (cationic antimicrobial peptides [62] or mutations that create large channels in the outer membrane [63]) make cells more susceptible to antibiotics.

The outer membrane contains proteins (porins) which form water-filled channels for diffusion of hydrophilic molecules. Porins play an important physiological role in the transport of various compounds. β-lactams, aminoglycosides, tetracyclines, and some fluoroquinolones can pass through porin channels [64, 65]. The loss of these porin channels can decrease the susceptibility of *P. aeruginosa* to antibiotics. Approximately 163 known or predicted outer membrane proteins were identified with 64 of these outer membrane proteins grouped into three families of porins [66]. OprF is a major porin of *P. aeruginosa* that forms a majority of small channels and a minority of larger channels [67]. OprF is present in high abundance as a closed conformer, and exists as an open channel only at very low levels. Therefore, it was shown that resistance to β-lactam antibiotics does not seem to involve loss or modification of OprF [68].

Porin OprD takes part in uptake of basic amino acids, small peptides and carbapenems (such as imipenem and meropenem) [69, 70]. Any substitution or deletion within external loop 2 and loop 3 of OprD results in changes of conformation and can cause imipenem resistance. Functional deletion of loop 2 at H729 induced partial resistance to imipenem and meropenem. Imipenam was found to bind to sites in loop 2 to block channel function. Deletion of loops 3 and 4 in OprD also results in failed expression. However, loop 3 is more likely to serve as a passage channel within OprD for imipenem, but not a direct binding site. Loop 1, loop 5, loop 6, loop 7, and loop 8 are not involved in the passage of imipenem, but either the deletion or amino acid substitutions of loop 5, loop 7, and loop 8 resulted in increased susceptibility to β-lactams; quinolones, chloramphenicol, carbapenems and tetracycline [71-76]. Amino acids including histidine, arginine, and lysine, its analogs, and peptides containing lysine can inhibit the penetration of imipenem in *P. aeruginosa* cells [70]. Culture medium containing basic amino acids significantly increased the minimum inhibitory concentration (MIC) of carbapenems against clinical isolates of *P. aeruginosa* [77].

Polycationic antibiotics, such as polymyxin B and aminoglycosides, and EDTA can pass through outer membrane without porins [78]. They displace divalent cations from lipopolysaccharide (LPS) molecules and destabilize the outer membrane increasing susceptibility to these antibiotics [79, 80]. Overexpression of OprH as a result of mutation or adaptation to low Mg^{2+} concentrations increases membrane resistance. OprH binds to LPS sites which are occupied by divalent cations and prevents access of polymyxin, gentamicin, and EDTA to these sites [78].

Besides porins, *P. aeruginosa* has numerous and highly efficient efflux mechanisms to resist to antibiotics. Efflux pumps include five superfamilies, based on energy source, the phylogenic relationship and the substrate specificity. There are five superfamilies: SMR (Small Multidrug Resistance), MET (Multidrug Endosomal Transporter), MAR (Multi Antimicrobial Resistance), RND (Resistance Nodulation Division), and MFS (Major Facilitator Superfamily) [81]. *P. aeruginosa* has efflux systems from all five superfamilies, but the largest number of predicted pumps belongs to the RND family with a total of 12 RND systems including two divalent metal cation transporters [82]. The efflux systems are composed of three protein components: an energy-dependent pump located in the cytoplasmic membrane, an outer membrane porin, and a linker protein which couples the two membrane components together. The 10 RND pumps of *P. aeruginosa* without the metal cation transporters are *MexAB-OprM*, *MexCD-OprJ*, *MexEF-OprN*, *MexXY*, *MexJK*, *MexGHI-OpmD*, *MexVW*, *MexPQ-OpmE*, *MexMN*, and *TriABC*, however, not all of these systems are well studied. These systems provide for intrinsic resistance to a number of antibiotics. Deletion, disruption or overexpression of pumps can make strains more or less sensitive to antibiotics or both effects can be shown (in case of *MexCD-oprJ*) [83].

Additionally, *P. aeruginosa* has a number of β-lactamases that are able to hydrolyze such antibiotics as penicillins, monobactams, cephalosporins, and carbapenems. β-lactamases divide into four classes, each including types that are usually plasmid-mediated or chromosomal [84]. The most common imported β-lactamases of *P. aeruginosa* are penicillinases from the molecular class A serine β-lactamases (PSE, CARB, and TEM families). The most prevalent enzymes of this group belong to the PSE family. Class A extended-spectrum β-lactamases also include enzymes from the TEM, SHV, CTX-M, PER, VEB, GES, and IBC families. Extended-spectrum β-lactamases from the class D, metallo-β-lactamases from the class B with four major families (IMP, VIM, SPM, and GIM families), OXA-type enzymes, class A carbapenemases of the KPC family also have been found within *P. aeruginosa*. *P. aeruginosa* has an inducible AmpC cephalosporinase which is similar to AmpC of several members of the *Enterobacteriaceae*. Increasing AmpC production provides for resistance to all β-lactams, except the carbapenems.

However, lack of AmpC increases susceptibility to imipenem and doripenem but not to meropenem. Overproduction of AmpC can occur either by induction of the ampC gene or through a process of derepression. Overproduction via induction occurs under the influence of specific β-lactams and β-lactamase inhibitors (cefoxitin, imipenem, and clavulanate), but the process is reversible after removal of the inducing agent. AmpC derepression is related to chromosomal mutations, and therefore concentration of AmpC enzyme remains at an elevated level [83].

Another mechanism of antibiotic resistance is modification of antibiotics such as aminoglycosides. Modifying enzymes phosphorylate (aminoglycoside phosphoryltransferase), acetylate (aminoglycoside acetyltransferase), or adenylate (aminoglycoside nucleotidyltransferase) these antibiotics. Aminoglycoside acetyltransferases (AAC) acetylate compounds such as gentamicin, tobramycin, netilmicin, and amikacin at the 1, 3, 6′, and 2′ amino groups. Aminoglycoside phosphoryltransferases (APH) inactivate kanamycin, neomycin, and streptomycin by modification of the 3′-OH of these antibiotics. Primary role of some phosphotransferases such as APH(3′)-IIb may be participation in metabolism, and resistance to aminoglycosides may be provided fortuitously. Aminoglycoside nucleotidyltransferases (ANT) modify aminoglycosides such as streptomycin and gentamicin. ANT(2")-I with AAC(6′) and AAC(3) are the most common enzymes providing for aminoglycoside resistance in *P. aeruginosa*. Enzymes that modify aminoglycosides can be associated with transposons which additionally carry genes for resistance to other compounds. *aac(3)* and *aac(6′)* genes are often associated with transposons or integrons carrying genes for extended-spectrum β-lactamases, metallo-β-lactamases or genes encoding other aminoglycoside-modifying enzymes [85].

Antibiotic resistance can be provided by changes in targets. Mutations in genes *gyrA* and *parC* (topoisomerases II and IV, respectively) increase resistance to fluoroquinolones. Mainly changes of *gyrA* especially in the Thr-83 codon provide reduced fluoroquinolone sensitivity in *P. aeruginosa*. Usually mutations in *parC* are found jointly with highly resistant *gyrA* mutants [86-88].

Biofilm-forming ability provides resistance to adverse conditions, like antibiotic tolerance in *P. aeruginosa*. Biofilm bacteria are usually embedded in an extracellular polymeric substance (EPS) matrix composed of polysaccharides, proteins, and nucleic acid [89-92]. The composition of the matrix depends on the environmental conditions, the age of the biofilm, and the particular *P. aeruginosa* strain forming the biofilm. At least three exopolysaccharides have been shown to be produced by *P. aeruginosa*: alginate, Psl, and Pel. Alginates are linear polyanionic exopolysaccharides composed of uronic acids. These compounds decrease susceptibility of biofilms to antibiotic treatment. The Psl polysaccharide is rich in mannose and galactose and is connected with initial attachment and mature biofilm formation. Pel is a glucose-rich, cellulose-like polymer that plays a role in cell-to-cell interactions [93]. Several mechanisms in biofilm increase resistance to antimicrobial agents. These are binding and sequestration of antimicrobial agents by EPS components, stationary phase or slow growth of cells because of nutrient and oxygen limitation within the depths of a biofilm [94, 95]. Alginate produced by *P. aeruginosa* can retard the diffusion of some antimicrobials (piperacillin, amikacin, gentamicin), whereas others penetrate readily (ciprofloxacin, levofloxacin, sparfloxacin, ofloxacin) [96,

97]. Addition of alginate lyase and DNase increase activity of antibiotics [98]. Biofilms are characterized by the heterogeneity: cells close to the substratum exhibit low metabolic activity and cells on top exhibit high metabolic activity. Antimicrobial agents such as ciprofloxacin, tetracycline, tobramycin, and gentamicin interfere with physiological processes of bacterial cells and specifically kill the metabolically active cells in the top layer of biofilms. Other antimicrobial agents such as colistin, EDTA, and SDS interfere with bacterial membrane structures and kill the cells of the deeper layer [99, 100]. However, a small number of bacteria can survive under simultaneous action of both treatments [99].

Thereby *P. aeruginosa* have many mechanisms allowing to survive negative effects of antibiotics. As a result *Pseudomonas* infections are hard to get rid of.

6. Waste as media for growth of *Pseudomonas bacteria*

As mentioned above, *Pseudomonas* can grow in minimal media and can utilize a large variety of organic molecules. It appears attractive to use waste as media for *Pseudomonas* cultivation, biodegradation or production of necessary compounds, hence further experiments were carried out.

Frying oil is produced in large quantities by the food industry and private households. The used cooking oil changes its composition and contains more than 30% of polar compounds depending on the variety of food, the type of frying and the number of cycles used. The utilization of these compounds is a growing problem, arousing expanding interest in the use of waste in microbial transformation [101]. Most of the tested *Pseudomonas* showed satisfactory growth on basal medium with 2% or 4% used olive oil or used sunflower oil. Used olive oil also induced biosurfactant production. Sunflower oil was worse substrate for cell growth and biosurfactant production [102].

Biosurfactants are the surface-active compounds that find use in the cosmetic and food production, healthcare, pulp and paper processing, coal, ceramic, and metal industries. They also may be applied in cleaning of oil-contaminated tankers, oil spill removal, transportation and recovery of crude oil, and bioremediation of contaminated sites. Biosurfactants show advantages over chemical analogs owing to their low toxicity and biodegradable nature. *Pseudomonas* is able to synthesize these compounds from cheap carbon sources such as vegetable oils and wastes from the food industry [58, 103].

P. aeruginosa LBI strain was grown on media containing one of residues from soybean, corn, babassu, cottonseed, and palm oil refinery. The soybean soapstock waste was the preferred substrate generating 11.7 g/L of rhamnolipids with the best surface-active properties compared with the products from other oil wastes. Biosurfactant from palm oil waste shows a good emulsification index against kerosene suggesting its potential use for bioremediation [104].

Similar experiments showed that waste motor lubricant oil and peanut oil cake [105], waste frying rice bran oil [106], distillery and whey wastes [107], waste frying coconut oil [108], olive oil mill wastewater [109] and molasses [110] can be used as cheap carbon sources for production of biosurfactants by *Pseudomonas*. Additionally, these substrates may help solve waste disposal problem.

Glycerol, cassava wastewater, waste cooking oil and cassava wastewater with waste frying oils were evaluated as alternative low-cost carbon substrates for the production of rhamnolipids and polyhydroxyalkanoates (PHAs) by various *P. aeruginosa* strains. Cassava wastewater with added waste cooking oil provides higher levels of rhamnolipids and PHAs compared with the other carbon substrates [111].

PHAs are composed of medium-chain length (R)-3-hydroxyfatty acids characterized by thermoplastic properties, biodegradability and biocompatibility. They make PHAs suitable for use in the packaging, medicine, pharmacy, agriculture and food industries [58]. Technical oleic acid and waste frying oil were shown to be suitable substrates for PHAs production by *P. aeruginosa* strain NCIB 40045 [112]. Glycerol by-product generated during the production of biodiesel from kitchen chimney dump lard was a better carbon source for PHA synthesis by *P. aeruginosa* JQ866912 as compared with commercial glycerol, sugarcane molasses and glucose. Using this glycerol by-product as a carbon source for PHA production could be both environmentally benign and cost-effective coupling of biodiesel and PHA production [113]. *P. oleovorans* is able to produce PHAs using the residual oil from biotechnological rhamnose production as the sole carbon source. PHAs isolated from *P. oleovorans* are more diverse than PHAs from *Ralstonia eutropha* H16 growing under the same conditions [114]. *P. putida* KT2442 produces PHAs in wastewater from olive oil mills (called alpechín), supplemented with glucose, yeast extract and NH_4Cl [115].

Wastes can be used as media in melanin production. Melanins represent a group of macromolecules, synthesized in living organisms by oxidative polymerization of various phenolic substances in the process of adaption [116]. Melanins act as photoprotectants against UV and visible light, charge transport mediators, free-radical scavengers, antioxidants, metal ion balancers [117]. Melanins find applications in agriculture, medicine, cosmetic and pharmaceutical industries. Some bacteria are able to synthesize these compounds. Marine melanin producer *Pseudomonas sp.* (closely related to *P. guinea*) was incubated in marine broth, vegetable waste from cabbage leftovers supplemented with 1.9 % NaCl to maintain salinity and marine broth - vegetable waste medium blended in 30:70 ratio for melanin production. The sole vegetable waste generated no pigmentation. Marine broth medium demonstrated more melanin production than the marine broth - vegetable waste blended medium (5.35 ± 0.4 and 2.79 ± 0.2 mg/mL after 72h of incubation, respectively). However, melanin from both sources after purification looked alike in appearance. This study confirms that the pigment can be produced from the cheaper substrates without any functional variation [118].

Another possible waste substrate as fermentation media is animal fleshing, the solid waste produced in large amounts by tanning industry. The studied *P. aeruginosa* strain can digest the media and produce alkaline protease, an industrially important enzyme from waste material. Alkaline proteases have considerable application in leather tanning industry [119]. Strain showed maximum alkaline protease production after 20 hours of incubation at the end of exponential growth phase [120].

P. aeruginosa MN7 was found to produce proteases when it was grown in media containing only shrimp waste powder, indicating that it can obtain its carbon, nitrogen, and salts requirements directly from shrimp waste. Protease production increased with increasing concentration of shrimp waste powder and reached a maximum value at 60 g/L [121]. Shrimp shell powder can be used for low-cost production of chitinase and chitosanase showing potential applications in the biocontrol of plant pathogenic fungi and insects. Shrimp shell powder (10 mU/mL) was more suitable as an inducer of chitinase production than squid pen powder (7.2 mU/mL), shrimp and crab shell powder (2.8 mU/mL), katsuobushi from mackerel (<0.1 mU/mL), katsuobushi from bonito (<0.1 mU/mL), and chitin (<0.1 mU/mL) [122].

The potential use of keratinous and chitinous wastes, such as chicken-feathers and shrimp wastes for oil-remediation was shown. Cultures were grown in minimal media with crude oil, or oil supplemented with chicken-feathers or shrimp wastes. The presence of organic wastes, mainly keratinous ones, enhanced the oil-hydrocarbons removal to an extent of 90%. Keratinolytic bacteria were better enzyme producers than the chitinolytic ones, and oil removal in the presence of chicken-feathers was 3.8 times higher than with shrimp wastes, and almost twice, in comparison with oil-only added cultures [123].

Various combinations of agricultural wastes can be tested to promote *P. fluorescence* production. Seven different variants were checked to detect the increased production of *P. fluorescence*. Composition containing rice straw, rice husk, wheat husk, cow dung, coconut water was found to be the optimal substrate for cultivation. The chosen combination also favored a high rate of green pigment production in this medium [124].

Toner waste black powder (TWBP) from copiers and printers is considered to be toxic for environment, and introduction of bacteria can alleviate the problem of TWBP disposal. It was stated that *P. spp.* and *P. aeuroginosa* utilize TWBP for growth. TWBP was mixed with soil at different concentrations (2g TWBP + 10g soil, 4g TWBP + 10 g soil, 6g TWBP + 10g soil, 8g TWBP + 10g soil and 10g TWBP + 10g soil) and inoculated in minimal salt medium. Among the various tested TWBP concentrations, 2g TWBP dose provoked significant stimulation of bacterial growth [125].

Tobacco-related processes can release wastes saturated with water-soluble nicotine posing biological and ecological hazard. *P. sp.* ZUTSKD consumed nicotine as sole source of carbon, nitrogen and energy when grown in basic inorganic salt medium. Growth and nicotine degradation were observed at substrate concentrations of 2–5.8 g/L. The strain degraded nicotine completely when the concentration of reducing sugar in TWE (tobacco waste extract) was lower than 8 g/L. Glucose concentration above 10 g/L inhibited nicotine degradation. Yeast extract and phosphate additions improved nicotine degradation in 5% TWE [126].

7. Stress resistance of *Pseudomonas* bacteria

Pseudomonas species thrive under moist conditions in soil (particularly in association with plants), in sewage sediments and the aquatic environments. Environmental conditions which will affect their growth include nutrient availability, moisture, temperature, competition, UV irradiation, oxygen availability, salinity and the presence of inhibitory or toxic compounds, but nutritional demands of *Pseudomonas* are modest [59].

There are some ways that allow pseudomonads to resist to adverse conditions. The alternative sigma factors RpoS (σ^s) and RpoE (σ^{22}; also referred to as AlgU or AlgT in fluorescent pseudomonads) are involved in bacterial survival under stress conditions. The sigma factor encoded by the *rpoS* gene is known to be important for survival under stressful conditions in several bacterial species. Studies of *rpoS* mutant *P. aeruginosa* PAOl revealed a two- to threefold increase in the rate of kill of stationary-phase cells following exposure to heat, low pH, high osmolarity, hydrogen peroxide and ethanol. However, stationary-phase *RpoS*-negative cells of *P. aeruginosa* were much more resistant than exponentially growing *RpoS*-positive cells [127]. *RpoS* gene also is involved in tolerance to antibiotics in *P. aeruginosa* during the stationary phase and heat stress [128].

The sigma factor AlgU contributes to tolerance towards osmotic, oxidative, and heat stresses in the pathogens *P. aeruginosa* and *P. syringae* [129-133]. AlgU in *P. aeruginosa* also plays part in regulation of biosynthesis of EPS alginate. AlgU is essential for adaptation of plant-associated *P. fluorescens* to osmotic and desiccation stresses [134]. *mucABCD* genes ensure tight control of AlgU activity [135]. The *mucA* gene encoding a transmembrane protein, and *mucB* gene encoding a periplasmic protein are negative regulators of AlgU. Stress conditions destabilize the MucB-MucA-AlgU complex, leading to release of AlgU into the cytosol where AlgU becomes active [136].

Production of some compounds can provide bacterial resistance to adverse conditions. PHA-negative mutants were more sensitive to heat treatment than non-mutated cells. The similar effect was revealed in biofilms of PHA-negative mutants as compared to non-mutated strains [137]. PHA availability enhances the ATP and ppGpp levels, and ppGpp has been shown to induce expression of the *rpoS* gene involved in regulation of stress tolerance [138].

P. putida NBAII-RPF9 can survive under saline shock (1 M NaCl for 1 h) or heat shock (45°C for 20 min). It was identified 13 upregulated proteins and one downregulated protein under heat shock, 6 upregulated proteins under heat tolerance, 11 upregulated proteins under saline shock, and 6 upregulated proteins under saline tolerance. During heat shock, heat stress responsive molecular chaperones and membrane proteins, and during salt stress, proteins upregulated to favor growth and adaptation of the bacterium were revealed. Heat shock chaperones DnaK and DnaJ were expressed under both saline and heat stress. The expression of different classes of proteins under abiotic stress can help this organism to adapt and survive under harsh environmental conditions [139].

Study of *P. aeruginosa* culture exposed to steady-state hyperosmotic stress demonstrated increased gene expression (at least threefold) in cells grown in the presence of 0.3 M NaCl and 0.7 M sucrose. Research revealed that 66 genes changed expression level in response to both stressors [140]. Also 40 of those 66 genes are associated with virulence factor expression, encoding proteins of a type III secretion system (TTSS), the type III cytotoxins ExoT and ExoY, and two ancillary chaperones [141, 142]. It has been shown that *P. aeruginosa* accumulated K^+, glutamate, trehalose as cytoplasmic osmoprotectants coupled to major organic osmoprotectant N-acetylglutaminylglutamine amide (NAGGN). Exogenous betaine was found to increase the growth rate and to partially replace NAGGN in osmotically stressed wild-type *P. aeruginosa* cells [143].

Organic solvents are extremely toxic to microbial cells, even at very low concentration. The cell membrane is the primary target for these compounds. Solvents penetrate into and disrupt the lipid bilayer of membrane. Concentration plays a crucial role in determining toxicity of organic solvents. Since Gram-negative bacteria have an additional outer membrane, and Gram-positive bacteria have a single cytoplasmic membrane, it was assumed that Gram-negative bacteria are better equipped to resist to organic solvents. Gram-negative bacteria including some strains of *Pseudomonas* possess various adaptive mechanisms of organic solvent tolerance. There are modifications in cell envelope to increase cell membrane rigidity and decrease permeability, enzymes increasing rate of membrane repair, special solvent-inactivating enzymes, action of efflux pumps, release of membrane vesicles with adhered solvent molecules. These mechanisms help bacteria to overcome the toxic effects of organic solvents [144].

As mentioned above, ability to form biofilm provides resistance to adverse conditions, like antibiotic exposure of *P. aeruginosa*. Biofilm beneficial impact is not limited exclusively to antibiotics. Biofilm cells were found to be more resistant to heavy metals than an equal number of free-floating cells. The degree of increased resistance varied depending on the element. EPS binds heavy metals and retards their diffusion within biofilm, protecting cells from stress [145].

8. Application of *Pseudomonas* bacteria

Due to simple requirements of growth conditions and medium composition, capacity to produce and degrade a number of compounds, *Pseudomonas* species are regarded as promising microorganisms in various biotechnological applications. As mentioned above, *Pseudomonas* is able to produce biosurfactants and PHAs characterized by low toxicity and biodegradability for further use in different technological areas. It's possible to apply waste in these processes as low-cost media.

Pseudomonas is also an excellent source of various enzymes acting as catalysts in specific biochemical reactions. High efficiency and specificity facilitate introduction of enzymes in diverse industrial processes. Enzymes produced by *Pseudomonas* species can be used in leather processing for dehairing of hides [119, 146], hydrolysis of oils to concentrate the derived fatty acids for medical purpose [147], production of monoacylglycerols and hydrocinnamic esters used in food, pharmaceutical and cosmetic industries [148, 149], manufacturing of detergents [150], production of biodiesel [151], remediation [152], etc.

Another possible application of *Pseudomonas* is bioremediation. *Pseudomonas* is able to remove various toxic pollutants from natural environment. Crude oil is known to alter physical and biochemical characteristics of soil. Petroleum contains numerous components including alkanes, aromatics, resins and asphaltenes. Action of some *Pseudomonas* cultures was shown to degrade constituents of crude oil, automobile oil effluent, and diesel fuel [153-155]. Moreover, pseudomonads can remove heavy metals released into the environment with industrial and domestic wastewaters. The studies proved that *Pseudomonas* strains are able to dispose of such metals as Cr, Cd, Mn, Fe, Cu, Ni, Pb from wastes [156-158]. Some species possess enormous potential for the detoxification of pollutants containing pesticides and phenols [157].

The textile industry makes extensive use of synthetic chemicals as dyes. A significant proportion of these dyes entering the surrounding media via wastewater is toxic to the environment and humans [159]. Dyes obstruct light penetration and oxygen transfer in water reservoirs. They retain stability and persistence in the environment for a long term [160]. Various physicochemical methods have been used for decolorization of dyes in wastewater, but these methods are distinguished by low efficiency, high cost, limited application scope, and production of recalcitrant wastes [161]. Application of bacteria can solve problems typical to physicochemical methods. It was shown that different *Pseudomonas* species efficiently decolorize and degrade dyes. It's possible to increase decolorization rates by changing cultural conditions. The optimum pH and temperature values for color removal are 7–9 and about 37°C, respectively. Immobilization, anaerobic conditions and addition of some compounds, like yeast extract, promote enhanced decolorization rate. Elevated concentrations of dyes and oxygen decelerate color removal [162-167].

Pseudomonas may be used as biocontrol agents that reduce disease severity and promote plant growth. They stimulate growth by several mechanisms. The bacteria can produce some compounds that inhibit spread of plant pathogens. These compounds are siderophores, hydrogen cyanide, pyrrolnitrin, phenazine, 2,4-diacetyl phloroglucinol and lytic enzymes (chitinase, β-1,3-glucanase). The inhibitors can act on pathogens directly like chitinase degrading the fungal cell wall or indirectly like siderophore that binds iron (III) ions in the environment and restrains access of pathogen to these ions. Additionally, *Pseudomonas* provides activation of induced systemic resistance (ISR) or systemic acquired resistance (SAR) in plants. Resistance reveals as oxidative burst which can lead to cell death and prevention of pathogen spreading, changes in cell wall composition, production of phytoalexins and PR proteins [168-173].

9. Conclusion

Genus *Pseudomonas* represents a diverse group of bacteria including a large number of species. On the one hand, *Pseudomonas* is characterized by ability to grow in minimal media without growth factors; on the other hand, these bacteria are able to produce and degrade a broad spectrum of compounds. *Pseudomonas* is equipped with several mechanisms allowing to resist and survive under adverse conditions. Currently the genus became the attractive object of intensive research.

Pseudomonas can be used in bioremediation allowing to degrade toxic compounds and solve problems concerning utilization of wastes hazardous for environment and humans. It was shown that 11 central and many different peripheral pathways provide for bacterial degradation of a whole range of compounds. *Pseudomonas* promising application is bioremediation of oil-contaminated environment. Crude oil causes changes of soil valuable properties such as fertility, water-holding and binding capacity, permeability, and bioremediation appears the best way to treat the oil contamination problem.

Pseudomonas species are capable of synthesizing both low-molecular-weight compounds (rhamnolipids, enzymes) and polymers (polyhydroxyalkanoates) that are often characterized by better properties than chemical analogs. Their potential usage is manufacturing cosmetics, food, oil refining, leather and paper processing, coal, ceramic, metal industries, agriculture, biodiesel production and medicine. Experiments revealed that agricultural and industrial wastes are suitable substrates for production of biosurfactants, polyhydroxyalkanoates, enzymes, melanin, etc. Application of these substrates will solve problems related to utilization of wastes.

Vast potential of pseudomonads as biocontrol agents was demonstrated. *Pseudomonas* decrease negative influence of plant pathogens by various ways. They can either produce compounds that directly affect pathogens or stimulate development of induced resistance in plants. Summing up, *Pseudomonas* species and their products find applications in various fields primarily because they are capable to utilize a wide range of organic and inorganic compounds.

The recent technological advances in the area of genomics and proteomics are now beginning to lay out important avenue of research focused on the role of *Pseudomonas* bacteria and the molecular mechanisms of their beneficial action.

Author details

Galina Novik*, Victoria Savich and Elena Kiseleva

*Address all correspondence to: galina_novik@mbio.bas-net.by

Collection of Microorganisms, Institute of Microbiology, Belarus National Academy of Sciences, Belarus

References

[1] Shinoda S, Okamoto K. Formation and function of Vibrio parahemolyticus lateral flagella. Journal of Bacteriology 1977;129(3) 1266-1271.

[2] Moore ERB, Tindall BJ, Martins Dos Santos VAP, Pieper DH, Ramos JL, Palleroni NJ. Nonmedical: Pseudomonas. In: Dworkin M, Falkow S, Rosenberg E., Schleifer KH, Stackebrandt E. (eds.) The Prokaryotes Volume 6: Proteobacteria: Gamma Subclass. Springer New York; 2006. p.646-703. DOI: 10.1007/0-387-30746-X_21.

[3] Woods DE, Straus DC, Johanson WG, Berry VK, Bass JA. Role of pili in adherence of Pseudomonas aeruginosa to mammalian buccal epithelial-cells. Infection and Immunity 1980;29(3) 1146-1151.

[4] Doig P, Todd T, Sastry PA, Lee KK, Hodges RS, Paranchych W, Irvin RT. Role of pili in adhesion of Pseudomonas aeruginosa to human respiratory epithelial-cells. Infection and Immunity 1988:56(6) 1641-1646.

[5] Bradley DE. A function of Pseudomonas aeruginosa PAO polar pili: twitching motility. Can Journal of Bacteriology 1980;26(2) 146-154. DOI: 10.1139/m80-022.

[6] Klausen M, Heydorn A, Ragas P, Lambertsen L, Aaes-Jørgensen A, Molin S, Tolker-Nielsen T. Biofilm formation by Pseudomonas aeruginosa wild type, flagella and type IV pili mutants. Molecular Microbiology 2003;48(6) 1511-1524. DOI: 10.1046/j.1365-2958.2003.03525.x.

[7] O'Toole GA, Kolter R. Flagellar and twitching motility are necessary for Pseudomonas aeruginosa biofilm development. Molecular Microbiology 1998;30(2) 295-304. DOI: 10.1046/j.1365-2958.1998.01062.x.

[8] Li CM, Brown I, Mansfield J, Stevens C, Boureau T, Romantschuk M, Taira S. The Hrp pilus of Pseudomonas syringae elongates from its tip and acts as a conduit for translocation of the effector protein HrpZ. The EMBO Journal 2002;21(8) 1909-1915. DOI: 10.1093/emboj/21.8.1909.

[9] Ghafoor A, Hay ID, Rehm BH. Role of exopolysaccharides in Pseudomonas aeruginosa biofilm formation and architecture. Applied and Environmental Microbiology 2011;77(15) 5238-5246. DOI: 10.1128/AEM.00637-11.

[10] Meyer JM, Abdallah MA. The fluorescent pigment of Pseudomonas fluorescens: biosynthesis, purification and physicochemical properties. Journal of General Microbiology 1978;107(2) 319-328. DOI: 10.1099/00221287-107-2-319.

[11] Meyer JM, Hornsperger JM. Role of pyoverdine Pf, the iron binding fluorescent pigment of Pseudomonas fluorescens in iron transport. Journal of General Microbiology 1978;107(2) 329-331. DOI: 10.1099/00221287-107-2-329.

[12] Takase H, Nitanai H, Hoshino K, Otani T. Impact of siderophore production on Pseudomonas aeruginosa infections in immunosuppressed mice. Infection and Immunity 2000;68(4) 1834-1839. DOI: 10.1128/IAI.68.4.1834-1839.2000.

[13] Meyer JM, Geoffroy VA, Baida N, Gardan L, Izard D, Lemanceau P, Achouak W, Palleroni NJ. Siderophore typing, a powerful tool for the identification of fluorescent and nonfluorescent pseudomonads. Applied and Environmental Microbiology 2002;68(6) 2745-2753. DOI: 10.1128/AEM.68.6.2745-2753.2002.

[14] Palleroni NJ. Pseudomonas. In: Brenner DJ, Krieg NR, Staley JT, Garrity GM. (eds.) Bergey's Manual of Systematic Bacteriology, Volume Two the Proteobacteria Part B

the Gammaproteobacteria. 2nd ed. Springer US; 2005. p.323-379. DOI: 10.1007/0-387-28022-7.

[15] Blondel-Hill E, Henry DA, David P. Pseudomonas. In: Murray PR, Baron EJ, Jorgensen JH, Landry ML, Pfaller MA. (eds.) Manual of Clinical Microbiology. 9th ed. Washington, DC: ASM Press. 2007; p.734-748.

[16] Wolterink AFWM., Jonker AB, Kengen SWM, Stams AJM. Pseudomonas chloritidismutans sp. nov., a non-denitrifying, chlorate-reducing bacterium. International Journal of Systematic and Evolutionary Microbiology 2002;52(Pt 6) 2183-2190. DOI: 10.1099/ijs.0.02102-0.

[17] Lessie TG, Phibbs PV Jr. Alternative pathways of carbohydrate utilization in pseudomonads. Annual Review of Microbiology 1984;38: 359-388. DOI: 10.1146/annurev.mi.38.100184.002043.

[18] Jiménez JI, Nogales J, García JL, Díaz E. A genomic view of the catabolism of aromatic compounds in Pseudomonas. In: Timmis KN. (ed.) Handbook of Hydrocarbon and Lipid Microbiology. Springer Berlin Heidelberg; 2010. p.1297-1325. DOI: 10.1007/978-3-540-77587-4_91.

[19] Jiménez JI, Miñambres B, García JL, Díaz E. Genomic insights in the metabolism of aromatic compounds in Pseudomonas. In: Ramos JL. (ed.) Pseudomonas, Volume 3 Biosynthesis of Macromolecules and Molecular Metabolism. Springer US; 2004. p. 425-462. DOI: 10.1007/978-1-4419-9088-4_15.

[20] List of prokaryotic names with standing in nomenclature. http://www.bacterio.net (accessed 30 January 2015).

[21] Stanier RY, Palleroni NJ, Doudoroff M. The aerobic pseudomonads: a taxonomic study. Journal of General Microbiology 1966;43(2) 159-271. DOI: 10.1099/00221287-43-2-159.

[22] Palleroni NJ. The Pseudomonas story. Environmental Microbiology 2010;12(6) 1377-1383. DOI: 10.1111/j.1462-2920.2009.02041.x.

[23] Goris J, Konstantinidis KT, Klappenbach JA, Coenye T, Vandamme P, Tiedje JM. DNA-DNA hybridization values and their relationship to whole-genome sequence similarities. International Journal of Systematic and Evolutionary Microbiology 2007;57(Pt 1) 81-91. DOI: 10.1099/ijs.0.64483-0.

[24] Doi RH, Igarashi RT. Conservation of ribosomal and messenger ribonucleic acid cistrons in Bacillus species. Journal of Bacteriology 1965; 90(2) 384-390.

[25] Dubnau D, Smith I, Porell P, Marmur J. Gene conservation in Bacillus species. I. Conserved genetic and nucleic acid base sequence homologies. Proceedings of the National Academy of Sciences of the United States of America 1965;54(2) 491-498. DOI: 10.2307/72742.

[26] Palleroni NJ, Kunisawa R, Contopoulou R, Doudoroff M. Nucleic acid homologies in the genus Pseudomonas. International Journal of Systematic Bacteriology 1973;23(4) 333-339. DOI: 10.1099/00207713-23-4-333.

[27] Yabuuchi E, Kosako Y, Oyaizu H, Yano I, Hotta H, Hashimoto Y, Ezaki T, Arakawa M. Proposal of Burkholderia gen. nov. and transfer of seven species of the genus Pseudomonas homology group II to the new genus, with the type species Burkholderia cepacia (Palleroni and Holmes 1981) comb. nov. Microbiology and Immunology 1992;36(12) 1251-1275. DOI: 10.1111/j.1348-0421.1992.tb02129.x.

[28] Yabuuchi E, Kosako Y, Yano I, Hotta I, Nishiuchi Y. Transfer of two Burkholderia and an Alcaligenes species to Ralstonia gen. nov.: proposal of Ralstonia pickettii (Ralston, Palleroni and Doudoroff 1973) comb. nov., Ralstonia solanacearum (Smith 1896) comb. nov. and Ralstonia eutropha (Davis 1969) comb. nov. Microbiology and Immunology 1995;39(11) 897-904. DOI: 10.1111/j.1348-0421.1995.tb03275.x.

[29] Willems A, Busse J, Goor M, Pot B, Falsen E, Jantzen E, Hoste B, Gillis M, Kersters K, Auling G, De Ley J. Hydrogenophaga, a new genus of hydrogen-oxidizing bacteria that includes Hydrogenophaga flava comb. nov. (formerly Pseudomonas flava), Hydrogenophaga palleroni (formerly Pseudomonas palleroni), Hydrogenophaga pseudoflava (formerly Pseudomonas pseudoflava and ''Pseudomonas carboxydoflava''), and Hydrogenophaga taeniospiralis (formerly Pseudomonas taeniospiralis). International Journal of Systematic Bacteriology 1989;39(3) 319-333. DOI: 10.1099/00207713-39-3-319.

[30] Willems A, Falsen E, Pot B, Jantzen E, Hoste B, Vandamme P, Gillis M, Kersters K, De Ley J. Acidovorax, a new genus for Pseudomonas facilis, Pseudomonas delafieldii E. Falsen (EF) group 13, EF group 16, and several clinical isolates, with the species Acidovorax facilis comb. nov., Acidovorax delafieldeii comb. nov., and Acidovorax temperans sp. nov. International Journal of Systematic Bacteriology 1990;40(4) 384-398. DOI: 10.1099/00207713-40-4-384.

[31] Tamaoka J, Ha D-M, Komagata K. Reclassification of Pseudomonas acidovorans den Dooren de Jong 1926 and Pseudomonas testosteroni Marcus and Talalay 1956 as Comamonas acidovorans comb. nov. and Comamonas testosterone comb. nov., with an emended description of the genus Comamonas. International Journal of Systematic Bacteriology 1987;37(1) 52-59. DOI: 10.1099/00207713-37-1-52.

[32] Segers P, Vancanneyt M, Pot B, Torck U, Hoste B, Dewettinck D, Felsen E, Kersters K, De Vos P. Classification of Pseudomonas diminuta Leifson and Hugh 1954 and Pseudomonas vesicularis Büsing, Döll, and Freytag 1953 in Brevundimonas gen. nov. as Brevundimonas diminuta comb. nov. and Brevundimonas vesicularis comb. nov., respectively. International Journal of Systematic Bacteriology 1994;44(3) 499-510. DOI: 10.1099/00207713-44-3-499.

[33] Palleroni NJ, Bradbury JF. Stenotrophomonas, a new bacterial genus for Xanthomonas maltophilia (Hugh 1980) Swings et al. 1983. International Journal of Systematic Bacteriology 1993;43(3) 606-609. DOI: 10.1099/00207713-43-3-606.

[34] Patel JB. 16S rRNA gene sequencing for bacterial pathogen identification in the clinical laboratory. Molecular Diagnosis 2001;6(4) 313-321. DOI: 10.1007/BF03262067.

[35] Moore ERB, Mau M, Arnscheidt A, Böttger EC, Hutson RA, Collins MD, van de Peer Y, De Wachter R, Timmis KN. The determination and comparison of the 16S rRNA gene sequences of species of the genus Pseudomonas (sensu stricto) and estimation of the natural intrageneric relationships. Systematic and Applied Microbiology 1996;19(4) 478-492. DOI: 10.1016/S0723-2020(96)80021-X.

[36] Anzai Y, Kim H, Park JY, Wakabayashi H, Oyaizu H. Phylogenetic affiliation of the pseudomonads based on 16S rRNA sequence. International Journal of Systematic and Evolutionary Microbiology 2000;50(Pt 4) 1563-1589. DOI: 10.1099/00207713-50-4-1563.

[37] Janda JM, Abbott SL. 16S rRNA gene sequencing for bacterial identification in the diagnostic laboratory: pluses, perils, and pitfalls. Journal of Clinical Microbiology 2007;45(9) 2761-2764. DOI: 10.1128/JCM.01228-07.

[38] Yamamoto S, Harayama S. Phylogenetic relationships of Pseudomonas putida strains deduced from the nucleotide sequences of gyrB, rpoD and 16S rRNA genes. International Journal of Systematic Bacteriology 1998;48(Pt 3) 813-819. DOI: 10.1099/00207713-48-3-813.

[39] Yamamoto S, Kasai H, Arnold DL, Jackson RW, Vivian A, Harayama S. Phylogeny of the genus Pseudomonas: intrageneric structure reconstructed from the nucleotide sequences of gyrB and rpoD genes. Microbiology 2000;146 (Pt 10) 2385-2394.

[40] De Vos D, Bouton C, Sarniguet A, De Vos P, Vauterin M, Cornelis P. Sequence diversity of the oprI gene, coding for major outer membrane lipoprotein I, among rRNA group I pseudomonads. Journal of Bacteriology 1998;180(24) 6551-6556.

[41] Ait Tayeb L, Ageron E, Grimont F, Grimont PAD. Molecular phylogeny of the genus Pseudomonas based on rpoB sequences and application for the identification of isolates. Research in Microbiology 2005;156(5-6) 763-773. DOI: 10.1016/j.resmic.2005.02.009.

[42] Hilario E, Buckley TR, Young JM. Improved resolution on the phylogenetic relationships among Pseudomonas by the combined analysis of atp D, car A, rec A and 16S rDNA. Antonie Van Leeuwenhoek. 2004;86(1) 51-64. DOI: 10.1023/B:ANTO.0000024910.57117.16.

[43] Bodilis J, Barray S. Molecular evolution of the major outer-membrane protein gene (oprF) of Pseudomonas. Microbiology 2006;152(Pt 4) 1075-1088. DOI: 10.1099/mic.0.28656-0.

[44] Vinatzer BA, Bull CT. The impact of genomic approaches on our understanding of diversity and taxonomy of plant pathogenic bacteria. In: Jackson RW. (ed.) Plant Pathogenic Bacteria: Genomics and Molecular Biology. Horizon Scientific Press; 2009. p.37-61.

[45] Mulet M, Lalucat J, García-Valdés E. DNA sequence-based analysis of the Pseudomonas species. Environmental Microbiology 2010;12(6) 1513-1530. DOI: 10.1111/j.1462-2920.2010.02181.x.

[46] Guasp C, Moore ER, Lalucat J, Bennasar A. Utility of internally transcribed 16S-23S rDNA spacer regions for the definition of Pseudomonas stutzeri genomovars and other Pseudomonas species. International Journal of Systematic and Evolutionary Microbiology 2000;50(Pt 4) 1629-1639. DOI: 10.1099/00207713-50-4-1629.

[47] Scarpellini M, Franzetti L, Galli A. Development of PCR assay to identify Pseudomonas fluorescens and its biotype. FEMS Microbiology Letters 2004;236(2) 257-260. DOI: 10.1111/j.1574-6968.2004.tb09655.x.

[48] Jaturapahu T., Puttinaowarat S., Somsiri T. Detection and identification of Pseudomonas Spp. by polymerase chain reaction-reverse cross-blot hybridization (PCR-RCBH) with 16S-23S ribosomal RNA intergenic spacer probes. In: Walker P, Lester R, Bondad-Reantaso MG. (eds.) Diseases in Asian Aquaculture: Proceedings of the Fifth symposium on Diseases in Asian Aquaculture, 24-28 November 2002, Queensland, Australia. Fish Health Section, Asian Fisheries Society, Manila, Philippines. 2005. p. 447-456.

[49] Geider K. Differentiation and identification of Pseudomonas syringae pathovars by PCR- and PFGE-Analyses. In: Rudolph K, Burr TJ, Mansfield JW, Stead D, Vivian A, von Kietzell J. (eds.) Developments in Plant Pathology, Volume 9: Pseudomonas syringae Pathovars and Related Pathogens. Springer Netherlands; 1997. p.459-464. DOI: 10.1007/978-94-011-5472-7_82.

[50] Spencker FB, Haupt S, Claros MC, Walter S, Lietz T, Schille R, Rodloff AC. Epidemiologic characterization of Pseudomonas aeruginosa in patients with cystic fibrosis. Clinical Microbiology and Infection 2000;6(11) 600-607. DOI: 10.1046/j.1469-0691.2000.00171.x.

[51] Fothergill JL, White J, Foweraker JE, Walshaw MJ, Ledson MJ, Mahenthiralingam E, Winstanley C. Impact of Pseudomonas aeruginosa genomic instability on the application of typing methods for chronic cystic fibrosis infections. Journal of Clinical Microbiology 2010;48(6) 2053-2059. DOI: 10.1128/JCM.00019-10.

[52] Wolska K, Szweda P. A comparative evaluation of PCR ribotyping and ERIC PCR for determining the diversity of clinical Pseudomonas aeruginosa isolates. Polish Journal of Microbiology 2008;57(2) 157-163.

[53] Syrmis MW, O'Carroll MR, Sloots TP, Coulter C, Wainwright CE, Bell SC, Nissen MD. Rapid genotyping of Pseudomonas aeruginosa isolates harboured by adult and

paediatric patients with cystic fibrosis using repetitive-element-based PCR assays. Journal of Medical Microbiology 2004;53(Pt 11): 1089-1096. DOI: 10.1099/jmm.0.45611-0.

[54] Han MM, Mu LZ, Liu XP, Zhao J, Liu XF, Liu H. ERIC-PCR genotyping of Pseudomonas aeruginosa isolates from haemorrhagic pneumonia cases in mink. Veterinary Record Open 2014;1(1): e000043. DOI: 10.1136/vropen-2014-000043.

[55] Meyer JM. Siderotyping and bacterial taxonomy: a siderophore bank for a rapid identification at the species level of fluorescent and non-fluorescent Pseudomonas. In: Varma A, Chincholkar SB. (eds.) Soil Biology, Microbial Siderophores, Volume 12. Springer Berlin Heidelberg; 2007. p.43-66. DOI: 10.1007/978-3-540-71160-5_2.

[56] Meyer JM, Gruffaz C, Raharinosy V, Bezverbnaya I, Schäfer M, Budzikiewicz H. Siderotyping of fluorescent Pseudomonas: molecular mass determination by mass spectrometry as a powerful pyoverdine siderotyping method. Biometals 2008;21(3) 259-271. DOI: 10.1007/s10534-007-9115-6.

[57] Tourkya B, Boubellouta T, Dufour E, Leriche F. Fluorescence spectroscopy as a promising tool for a polyphasic approach to pseudomonad taxonomy. Current Microbiology 2009;58(1) 39-46. DOI: 10.1007/s00284-008-9263-0.

[58] Rehm BHA. Biotechnological relevance of Pseudomonads. In: Rehm BHA. (ed.) Pseudomonas: Model Organism, Pathogen, Cell Factory. Wiley-Blackwell; 2008. p. 377-395. DOI: 10.1002/9783527622009.ch14.

[59] OECD. Section 2 – Pseudomonas. In: Safety Assessment of Transgenic Organisms, Volume 2: OECD Consensus Documents. OECD Publishing; 2006. p.312-393. DOI: 10.1787/9789264095403-10-en.

[60] Choi JY, Sifri CD, Goumnerov BC, Rahme LG, Ausubel FM, Calderwood SB. Identification of virulence genes in a pathogenic strain of Pseudomonas aeruginosa by representational difference analysis. Journal of Bacteriology 2002;184(4): 952-961. DOI: 10.1128/jb.184.4.952-961.2002.

[61] Lambert PA. Mechanisms of antibiotic resistance in Pseudomonas aeruginosa. Journal of the Royal Society of Medicine 2002;95(Suppl 41) 22-26.

[62] Scott MG, Yan H, Hancock REW. Biological properties of structurally related α-helical cationic antimicrobial peptides. Infection and Immunity 1999;67(4) 2005-2009.

[63] Huang H, Hancock REW. The role of specific surface loop regions in determining the function of the imipenem-specific pore protein OprD of Pseudomonas aeruginosa. Journal of Bacteriology 1996;178(11) 3085-3090.

[64] Nikaido H. Outer membrane barrier as a mechanism of antimicrobial resistance. Antimicrobial Agents and Chemotherapy 1989;33(11): 1831-1836. DOI: 10.1128/AAC.33.11.1831.

[65] Yoshimura F, Nikaido H. Diffusion of beta-lactam antibiotics through the porin channels of Escherichia coli K-12. Antimicrobial Agents and Chemotherapy 1985;27(1) 84-92. DOI: 10.1128/AAC.27.1.84.

[66] Hancock REW, Brinkman FS. Function of pseudomonas porins in uptake and efflux. Annual Review of Microbiology 2002;56 17-38. DOI: 10.1146/annurev.micro.56.012302.160310.

[67] Brinkman FS, Bains M, Hancock RE. The amino terminus of Pseudomonas aeruginosa outer membrane protein OprF forms channels in lipid bilayer membranes: correlation with a three-dimensional model. Journal of Bacteriology 2000;182(18) 5251-5255. DOI: 10.1128/JB.182.18.5251-5255.2000.

[68] Bratu S, Landman D, Gupta J, Quale J. Role of AmpD, OprF and penicillin-binding proteins in beta-lactam resistance in clinical isolates of Pseudomonas aeruginosa. Journal of Medical Microbiology 2007;56(6) 809-814. DOI: 10.1099/jmm.0.47019-0.

[69] Trias J, Nikaido H. Outer membrane protein D2 catalyzes facilitated diffusion of carbapenems and penems through the outer membrane of Pseudomonas aeruginosa. Antimicrobial Agents and Chemotherapy 1990;34(1) 52-57. DOI: 10.1128/AAC.34.1.52.

[70] Trias J, Nikaido H. Protein D2 channel of the Pseudomonas aeruginosa outer membrane has a binding site for basic amino acids and peptides. Journal of Biological Chemistry 1990;265(26) 15680-15684.

[71] Ochs MM, Bains M, Hancock REW. Role of putative loops 2 and 3 in imipenem passage through the specific porin OprD of Pseudomonas aeruginosa. Antimicrobial Agents and Chemotherapy 2000;44(7) 1983-1985. DOI: 10.1128/AAC.44.7.1983-1985.2000.

[72] Huang H, Jeanteur D, Pattus F, Hancock REW. Membrane topology and site specific mutagenesis of Pseudomonas aeruginosa porin OprD. Molecular Microbiology 1995;16(5) 931–941. DOI: 10.1111/j.1365-2958.1995.tb02319.x.

[73] Huang H, Hancock RW. The role of specific surface loop regions in determining the function of the imipenem-specific pore protein OprD of Pseudomonas aeruginosa. Journal of Bacteriology 1996;178(11) 3085-3090.

[74] Pirnay JP, De Vos D, Mossialos D, Vanderkelen A, Cornelis P, Zizi M. Analysis of the Pseudomonas aeruginosa oprD gene from clinical and environmental isolates. Environmental Microbiology 2002;4(12) 872-882. DOI: 10.1046/j.1462-2920.2002.00281.x.

[75] Chevalier S, Bodilis J, Jaouen T, Barray S, Feuilloley MGJ, Orange N. Sequence diversity of the OprD protein of environmental Pseudomonas strains. Environmental Microbiology 2007;9(3) 824-835. DOI: 10.1111/j.1462-2920.2006.01191.x.

[76] Epp SF, Köhler T, Plésiat P, Michéa-Hamzehpour M, Frey J, Pechère JC. C-terminal region of Pseudomonas aeruginosa outer membrane porin OprD modulates suscepti-

bility to meropenem. Antimicrobial Agents and Chemotherapy 2001;45(6) 1780-1787. DOI: 10.1128/AAC.45.6.1780-1787.2001.

[77] Muramatsu H, Horii T, Morita M, Hashimoto H, Kanno T, Maekawa M. Effect of basic amino acids on susceptibility to carbapenems in clinical Pseudomonas aeruginosa isolates. International Journal of Medical Microbiology 2003;293(2-3) 191-197. DOI: 10.1078/1438-4221-00256.

[78] Young ML, Bains M, Bell A, Hancock RE. Role of Pseudomonas aeruginosa outer membrane protein OprH in polymyxin and gentamicin resistance: isolation of an OprH-deficient mutant by gene replacement techniques. Antimicrobial Agents and Chemotherapy 1992;36(11) 2566-2568. DOI: 10.1128/AAC.36.11.2566.

[79] Hancock RE, Chan L. Outer membranes of environmental isolates of Pseudomonas aeruginosa. Journal of Clinical Microbiology 1988;26(11) 2423-2424.

[80] M Vaara. Agents that increase the permeability of the outer membrane. Microbiological Reviews 1992;56(3) 395-411.

[81] Van Bambeke R, Balzi E, Tulkens PM. Antibiotic efflux pumps. Biochemical Pharmacology 2000;60(4) 457-470. DOI: 10.1016/S0006-2952(00)00291-4.

[82] Stover CK, Pham XQ, Erwin AL, Mizoguchi SD, Warrener P, Hickey MJ, Brinkman FS, Hufnagle WO, Kowalik DJ, Lagrou M, Garber RL, Goltry L, Tolentino E, Westbrock-Wadman S, Yuan Y, Brody LL, Coulter SN, Folger KR, Kas A, Larbig K, Lim R, Smith K, Spencer D, Wong GK, Wu Z, Paulsen IT, Reizer J, Saier MH, Hancock RE, Lory S, Olson MV. Complete genome sequence of Pseudomonas aeruginosa PAO1, an opportunistic pathogen. Nature 2000;406(6799) 959-964. DOI: 10.1038/35023079.

[83] Lister PD, Wolter DJ, Hanson ND. Antibacterial-resistant Pseudomonas aeruginosa: clinical impact and complex regulation of chromosomally encoded resistance mechanisms. Clinical Microbiology Reviews 2009;22(4) 582-610. DOI: 10.1128/CMR.00040-09.

[84] Livermore DM, Woodford N. The beta-lactamase threat in Enterobacteriaceae, Pseudomonas and Acinetobacter. Trends in Microbiology 2006;14(9) 413-420. DOI: 10.1016/j.tim.2006.07.008.

[85] Poole K. Aminoglycoside resistance in Pseudomonas aeruginosa. Antimicrobial Agents and Chemotherapy 2005;49(2) 479-487. DOI: 10.1128/AAC.49.2.479-487.2005.

[86] Yonezawa M, Takahata M, Matsubara N, Watanabe Y, Narita H. DNA gyrase gyrA mutations in quinolone-resistant clinical isolates of Pseudomonas aeruginosa. Antimicrobial Agents and Chemotherapy 1995;39(9) 1970-1972. DOI: 10.1128/AAC.39.9.1970.

[87] Higgins PG, Fluit AC, Milatovic D, Verhoef J, Schmitz FJ. Mutations in GyrA, ParC, MexR and NfxB in clinical isolates of Pseudomonas aeruginosa. The International

Journal of Antimicrobial Agents 2003;21(5) 409-413. DOI: 10.1016/S0924-8579(03)00009-8.

[88] Salma R, Dabboussi F, Kassaa I, Khudary R, Hamze M. gyrA and parC mutations in quinolone-resistant clinical isolates of Pseudomonas aeruginosa from Nini Hospital in north Lebanon. Journal of Infection and Chemotherapy 2013;19(1) 77-81. DOI: 10.1007/s10156-012-0455-y.

[89] Flemming HC, Wingender J. Relevance of microbial extracellular polymeric substances (EPSs)-Part I: Structural and ecological aspects. Water Science and Technology 2001;43(6) 1-8.

[90] Sutherland IW. Exopolysaccharides in biofilms, flocs and related structures. Water Science and Technology 2001;43(6) 77-86.

[91] Whitchurch CB, Tolker-Nielsen T, Ragas PC, Mattick JS. Extracellular DNA required for bacterial biofilm formation. Science 2002;295(5559) 1487. DOI: 10.1126/science.295.5559.1487.

[92] Toyofuku M, Roschitzki B, Riedel K, Eberl L. Identification of proteins associated with the Pseudomonas aeruginosa biofilm extracellular matrix. Journal of Proteome Research 2012;11(10) 4906-4915. DOI: 10.1021/pr300395j.

[93] Ghafoor A, Hay ID, Rehm BH. Role of exopolysaccharides in Pseudomonas aeruginosa biofilm formation and architecture. Applied and Environmental Microbiology 2011;77(15) 5238-5246. DOI: 10.1128/AEM.00637-11.

[94] Borriello G, Werner E, Roe F, Kim AM, Ehrlich GD, Stewart PS. Oxygen limitation contributes to antibiotic tolerance of Pseudomonas aeruginosa in biofilms. Antimicrobial Agents and Chemotherapy 2004;48(7) 2659-2664. DOI: 10.1128/AAC.48.7.2659-2664.2004.

[95] Brown MRW, Allison DG, Gilbert P. Resistance of bacterial biofilms to antibiotics: a growth-rate related effect? Journal of Antimicrobial Chemotherapy 1988;22(6) 777-783. DOI: 10.1093/jac/22.6.777.

[96] Hoyle BD, Alcantara J, Costerton JW. Pseudomonas aeruginosa biofilm as a diffusion barrier to piperacillin. Antimicrobial Agents and Chemotherapy 1992;36(9) 2054-2056. DOI: 10.1128/AAC.36.9.2054.

[97] Shigeta M, Tanaka G, Komatsuzawa H, Sugai M, Suginaka H, Usui T. Permeation of antimicrobial agents through Pseudomonas aeruginosa biofilms: a simple method. Chemotherapy 1997;43(5) 340-345.

[98] Alipour M, Suntres ZE, Omri A. Importance of DNase and alginate lyase for enhancing free and liposome encapsulated aminoglycoside activity against Pseudomonas aeruginosa. Journal of Antimicrobial Chemotherapy 2009;64(2) 317-325. DOI: 10.1093/jac/dkp165.

[99] Pamp SJ, Gjermansen M, Johansen HK, Tolker-Nielsen T. Tolerance to the antimicrobial peptide colistin in Pseudomonas aeruginosa biofilms is linked to metabolically active cells, and depends on the pmr and mexAB-oprM genes. Molecular Microbiology 2008;68(1) 223-240. DOI: 10.1111/j.1365-2958.2008.06152.x.

[100] Banin E, Brady KM, Greenberg EP. Chelator-induced dispersal and killing of Pseudomonas aeruginosa cells in a biofilm. Applied and Environmental Microbiology 2006;72(3) 2064-2069. DOI: 10.1128/AEM.72.3.2064-2069.2006.

[101] Kock JLF, Botha A, Blerh J, Nigam S. Used cooking oil: science tackles a potential health hazard. South African Journal of Science 1996;92(11-12) 513–514.

[102] Haba E, Espuny MJ, Busquets M, Manresa A. Screening and production of rhamnolipids by Pseudomonas aeruginosa 47T2 NCIB 40044 from waste frying oils. Journal of Applied Microbiology 2000;88(3) 379-387. DOI: 10.1046/j.1365-2672.2000.00961.x.

[103] Makkar RS, Cameotra SS. An update on the use of unconventional substrates for biosurfactant production and their new applications. Applied Microbiology and Biotechnology 2002;58(4) 428-434. DOI: 10.1007/s00253-001-0924-1.

[104] Nitschke M, Costa SG, Haddad R, Gonçalves LA, Eberlin MN, Contiero J. Oil wastes as unconventional substrates for rhamnolipid biosurfactant production by Pseudomonas aeruginosa LBI. Biotechnology Progress 2005;21(5) 1562-1566. DOI: 10.1021/bp050198x.

[105] Thavasi R, Subramanyam Nambaru VRM, Jayalakshmi S, Balasubramanian T, Banat IM. Biosurfactant production by Pseudomonas aeruginosa from renewable resources. Indian Journal of Microbiology 2011;51(1) 30-36. DOI: 10.1007/s12088-011-0076-7.

[106] Venkatesh NM, Vedaraman N. Remediation of soil contaminated with copper using rhamnolipids produced from Pseudomonas aeruginosa MTCC 2297 using waste frying rice bran oil. Annals of Microbiology 2012;62(1) 85-91. DOI: 10.1007/s13213-011-0230-9.

[107] Babu PS, Vaidya AN, Bal AS, Kapur R, Juwarkar A, Khanna P. Kinetics of biosurfactant production by Pseudomonas aeruginosa strain BS2 from industrial wastes. Biotechnology Letters 1996;18(3) 263-268. DOI: 10.1007/BF00142942.

[108] George S, Jayachandran K. Production and characterization of rhamnolipid biosurfactant from waste frying coconut oil using a novel Pseudomonas aeruginosa D. Journal of Applied Microbiology 2013;114(2) 373-383. DOI: 10.1111/jam.12069.

[109] Colak AK, Kahraman H. The use of raw cheese whey and olive oil mill wastewater for rhamnolipid production by recombinant Pseudomonas aeruginosa. Environmental and Experimental Biology 2013;11(3) 125-130.

[110] Rashedi H, Assadi MM, Bonakdarpour B, Jamshidi E. Environmental importance of rhamnolipid production from molasses as a carbon source. Internation Journal of Environmental Science and Technology 2005;2(1) 59-62. DOI: 10.1007/BF03325858.

[111] Costa SG, Lépine F, Milot S, Déziel E, Nitschke M, Contiero J. Cassava wastewater as a substrate for the simultaneous production of rhamnolipids and polyhydroxyalkanoates by Pseudomonas aeruginosa. Journal of Industrial Microbiology and Biotechnology 2009;36(8) 1063-1072. DOI: 10.1007/s10295-009-0590-3.

[112] Fernández D, Rodríguez E, Bassas M, Viñas M, Solanas AM, Llorens J, Marqués AM, Manresa A. Agro-industrial oily wastes as substrates for PHA production by the new strain Pseudomonas aeruginosa NCIB 40045: Effect of culture conditions. Biochemical Engineering Journal 2005;26(2) 159-167. DOI: 10.1016/j.bej.2005.04.022.

[113] Phukon P, Phukan MM, Phukan S, Konwa BK. Polyhydroxyalkanoate production by indigenously isolated Pseudomonas aeruginosa using glycerol by-product of KCDL biodiesel as an inexpensive carbon source. Annals of Microbiology 2014;64(4) 1567-1574. DOI 10.1007/s13213-014-0800-8.

[114] Füchtenbusch B, Wullbrandt D, Steinbüchel A. Production of polyhydroxyalkanoic acids by Ralstonia eutropha and Pseudomonas oleovorans from an oil remaining from biotechnological rhamnose production. Applied Microbiology and Biotechnology 2000;53(2) 167-172. DOI: 10.1007/s002530050004.

[115] Ribera RG, Monteoliva-Sanchez M, Ramos-Cormenzana A. Production of polyhydroxyalkanoates by Pseudomonas putida KT2442 harboring pSK2665 in wastewater from olive oil mills (alpechín). Electronic Journal of Biotechnology 2001;4(2) 116-119.

[116] Sajjan S, Purification and physiochemical characterization of melanin pigment from Klebsiella sp. GSK. Journal of Microbiology and Biotechnology 2010;20(11) 1513-1520.

[117] Geng J, Yuan P, Shao C, Yu SB, Zhou B, Zhou P, Chen XD. Bacterial melanin interacts with double-stranded DNA with high affinity and may inhibit cell metabolism in vivo. Archives of Microbiology 2010;192(5) 321-329. DOI: 10.1007/s00203-010-0560-1.

[118] Tarangini K, Mishra S. Production, characterization and analysis of melanin from isolated marine Pseudomonas sp. using vegetable waste. Research Journal of Engineering Sciences 2013;2(5) 40-46.

[119] Dayanandan A, Kanagaraj J, Sounderraj L, Govindaraju R., Rajkumar GS. Application of an alkaline protease in leather processing: an ecofriendly approach. Journal of Cleaner Production 2003;11(5) 533-536. DOI: 10.1016/S0959-6526(02)00056-2.

[120] Kumar AG, Swarnalatha S, Sairam B, Sekaran G. Production of alkaline protease by Pseudomonas aeruginosa using proteinaceous solid waste generated from leather manufacturing industries. Bioresource Technology 2008;99(6) 1939-1944. DOI: 10.1016/j.biortech.2007.03.025.

[121] Jellouli K, Bayoudh A, Manni L, Agrebi R, Nasri M. Purification, biochemical and molecular characterization of a metalloprotease from Pseudomonas aeruginosa MN7

grown on shrimp wastes. Applied Microbiology and Biotechnology 2008;79(6) 989-999. DOI: 10.1007/s00253-008-1517-z.

[122] Wang SL, Chen SJ, Wang CL. Purification and characterization of chitinases and chitosanases from a new species strain Pseudomonas sp. TKU015 using shrimp shells as a substrate. Carbohydrate Research 2008;19;343(7) 1171-1179. DOI: 10.1016/j.carres.2008.03.018.

[123] Cervantes-González E, Rojas-Avelizapa NG, Cruz-Camarillo R, García-Mena J, Rojas-Avelizapa LI. Oil-removal enhancement in media with keratinous or chitinous wastes by hydrocarbon-degrading bacteria isolated from oil-polluted soils. Environmental Technology 2008;29(2) 171-182. DOI: 10.1080/09593330802028659.

[124] Poorni KE, Manikandan A, Geethanjali S, Percy PK. Production of Pseudomonas fluorescence from agricultural wastes and its application in the preservation of selected vegetables. Advances in Applied Science Research 2011;2(2) 156-160.

[125] Sepperumal U, Selvanayagam S, Markandan M. Utilization of toner waste black powder for bacterial growth. Journal of Microbiology and Biotechnology Research 2014;4(1) 28-30.

[126] Zhong W, Zhu C, Shu M, Sun K, Zhao L, Wanga C, Ye Z, Chen J. Degradation of nicotine in tobacco waste extract by newly isolated Pseudomonas sp. ZUTSKD. Bioresource Technology 2010;101(18) 6935-6941. DOI: 10.1016/j.biortech.2010.03.142.

[127] Jørgensen F, Bally M, Chapon-Herve V, Michel G, Lazdunski A, Williams P, Stewart GS. RpoS-dependent stress tolerance in Pseudomonas aeruginosa. Microbiology 1999;145(Pt 4) 835-844. DOI: 10.1099/13500872-145-4-835.

[128] Murakami K, Ono T, Viducic D, Kayama S, Mori M, Hirota K, Nemoto K, Miyake Y. Role for rpoS gene of Pseudomonas aeruginosa in antibiotic tolerance. FEMS Microbiology Letters 2005;242(1) 161-167. DOI: 10.1016/j.femsle.2004.11.005.

[129] Martin DW, Schurr MJ, Yu H, Deretic V. Analysis of promoters controlled by the putative sigma factor AlgU regulating conversion to mucoidy in Pseudomonas aeruginosa: relationship to σ^E and stress response. Journal of Bacteriology 1994;176(21) 6688-6696.

[130] Keith LMW, Bender CL. AlgT (σ^{22}) controls alginate production and tolerance to environmental stress in Pseudomonas syringae. Journal of Bacteriology 1999;181(23) 7176-7184.

[131] Schurr MJ, Deretic V. Microbial pathogenesis in cystic fibrosis: co-ordinate regulation of heat-shock response and conversion to mucoidy in Pseudomonas aeruginosa. Molecular Microbiology 1997;24(2) 411-420. DOI: 10.1046/j.1365-2958.1997.3411711.x.

[132] Schurr MJ, Yu H, Boucher JC, Hibler NS, Deretic V. Multiple promoters and induction by heat shock of the gene encoding the alternative sigma factor AlgU (σ^E) which

controls mucoidy in cystic fibrosis isolates of Pseudomonas aeruginosa. Journal of Bacteriology 1995;177(19) 5670-5679.

[133] Yu H, Schurr MJ, Deretic V. Functional equivalence of Escherichia coli σE and Pseudomonas aeruginosa AlgU: E. coli rpoE restores mucoidy and reduces sensitivity to reactive oxygen intermediates in algU mutants of P. aeruginosa. Journal of Bacteriology 1995;177(11) 3259-3268.

[134] Schnider-Keel U, Lejbølle KB, Baehler E, Haas D, Keel C. The sigma factor AlgU (AlgT) controls exopolysaccharide production and tolerance towards desiccation and osmotic stress in the biocontrol agent Pseudomonas fluorescens CHA0. Applied and Environmental Microbiology 2001;67(12) 5683-5693. DOI: 10.1128/AEM.67.12.5683-5693.2001.

[135] Govan J, Deretic V. Microbial pathogenesis in cystic fibrosis: mucoid Pseudomonas aeruginosa and Burkholderia cepacia. Microbiological Reviews 1996;60(3) 539-574.

[136] Mathee K, McPherson CJ, Ohman DE. Posttranslational control of the algT (algU)-encoded sigma22 for expression of the alginate regulon in Pseudomonas aeruginosa and localization of its antagonist proteins MucA and MucB (AlgN). Journal of Bacteriology 1997;179(11) 3711-3720.

[137] Pham TH, Webb JS, Rehm BHA. The role of polyhydroxyalkanoate biosynthesis by Pseudomonas aeruginosa in rhamnolipid and alginate production as well as stress tolerance and biofilm formation. Microbiology 2004;150(10) 3405-3413. DOI: 10.1099/mic.0.27357-0.

[138] Gentry DR, Hernandez VJ, Nguyen LH, Jensen DB, Cashel M. Synthesis of the stationary-phase sigma factor σ^s is positively regulated by ppGpp. Journal of Bacteriology 1993;175(24) 7982-7989.

[139] Rangeshwaran R, Ashwitha K, Sivakumar G, Jalali SK. Analysis of proteins expressed by an abiotic stress tolerant Pseudomonas putida (NBAII-RPF9) isolate under saline and high temperature conditions. Current Microbiology 2013;67(6) 659-667. DOI: 10.1007/s00284-013-0416-4.

[140] Aspedon A, Palmer K, Whiteley M. Microarray analysis of the osmotic stress response in Pseudomonas aeruginosa. Journal of Bacteriology 2006;188(7) 2721-2725. DOI: 10.1128/JB.188.7.2721-2725.2006.

[141] Frank DW. The exoenzyme S regulon of Pseudomonas aeruginosa. Molecular Microbiology 1997;26(4) 621-629. DOI: 10.1046/j.1365-2958.1997.6251991.x.

[142] Hueck CJ. Type III protein secretion systems in bacterial pathogens of animals and plants. Microbiology and Molecular Biology Reviews 1998;62(2) 379-433.

[143] D'Souza-Ault MR, Smith LT, Smith GM. Roles of N-acetylglutaminylglutamine amide and glycine betaine in adaptation of Pseudomonas aeruginosa to osmotic stress. Applied and Environmental Microbiology 1993;59(2) 473-478.

[144] Sardessai Y, Bhosle S. Tolerance of bacteria to organic solvents. Research in Microbiology 2002;153(5) 263-268. DOI: 10.1016/S0923-2508(02)01319-0.

[145] Teitzel GM, Parsek MR. Heavy metal resistance of biofilm and planktonic Pseudomonas aeruginosa. Applied and Environmental Microbiology 2003;69(4) 2313-2320. DOI: 10.1128/AEM.69.4.2313-2320.2003.

[146] Boopathy NR, Indhuja D, Srinivasan K, Uthirappan M, Gupta R, Ramudu KN, Chellan R. Statistical medium optimization of an alkaline protease from Pseudomonas aeruginosa MTCC 10501, its characterization and application in leather processing. Indian Journal of Experimental Biology 2013;51(4) 336-342.

[147] Kojima Y, Sakuradani E, Shimizu S. Different specificities of two types of Pseudomonas lipases for C20 fatty acids with a Δ5 unsaturated double bond and their application for selective concentration of fatty acids. Journal of Bioscience and Bioengineering 2006;101(6) 490-500. DOI: 10.1263/jbb.101.496.

[148] Cheirsilp B, Jeamjounkhaw P, Kittikun AH. Optimizing an alginate immobilized lipase for monoacylglycerol production by the glycerolysis reaction. Journal of Molecular Catalysis B: Enzymatic 2009;59(1-3) 206-211. DOI: 10.1016/j.molcatb.2009.03.001.

[149] Zhang A, Gao R, Diao N, Xie G, Gao G, Cao S. Cloning, expression and characterization of an organic solvent tolerant lipase from Pseudomonas fluorescens JCM5963. Journal of Molecular Catalysis B: Enzymatic 2009;56(2-3) 78-84. DOI: 10.1016/j.molcatb.2008.06.021.

[150] Priya K, Chadha A. Synthesis of hydrocinnamic esters by Pseudomonas cepacia lipase. Enzyme and Microbial Technology 2003;32(3-4) 885-890. DOI: 10.1016/S0141-0229(02)00340-X.

[151] Li Q, Yan Y. Production of biodiesel catalyzed by immobilized Pseudomonas cepacia lipase from Sapium sebiferum oil in micro-aqueous phase. Applied Energy 2008;87(10) 3148-3154. DOI: 03/13/201510.1016/j.apenergy.2010.02.032.

[152] Vandana P, Peter JK. Application of partially purified laccases from Pseudomonas fluorescens on dye decolourization. International Journal of Advanced Technology in Engineering and Science 2014;2(8) 317-327.

[153] Kumar V, Singh S, Manhas A, Singh J, Singla S, Kaur P, Data S, Negi P, Kalia A. Bioremediation of petroleum hydrocarbon by using Pseudomonas species isolated from petroleum contaminated soil. Oriental Journal of Chemistry 2014;30(4) 1771-1776.

[154] Sathiya-Moorthi P, Deecaraman M, Kalaichelvan PT. Bioremediation of automobile oil effluent by Pseudomonas sp. Advanced Biotech 2008;31 34-37.

[155] Roy AS, Yenn R, Singh AK, Boruah HPD, Saikia N, Deka M. Bioremediation of crude oil contaminated tea plantation soil using two Pseudomonas aeruginosa strains AS 03 and NA 108. African Journal of Biotechnology 2013;12(19) 2600-2610. DOI: 10.5897/AJB12.170

[156] Baltazar M, Gracioso L, Avanzi I, Veiga M, Gimenes L, Nascimento C, Perpetuo E. Bioremediation potential of Pseudomonas aeruginosa and Enterobacter cloacae isolated from a copper-contaminated area. BMC Proceedings 2014;8(Suppl 4) P188. DOI: 10.1186/1753-6561-8-S4-P188.

[157] Ali Khan MW, Ahmad M. Detoxification and bioremediation potential of a Pseudomonas fluorescens isolate against the major Indian water pollutants. Journal of Environmental Science and Health, Part A: Toxic/Hazardous Substances and Environmental Engineering 2006;41(4) 659-674. DOI: 10.1080/10934520600575051.

[158] Singh R, Bishnoi NR, Kirrolia A. Evaluation of Pseudomonas aeruginosa an innovative bioremediation tool in multi metals ions from simulated system using multi response methodology. Bioresource Technology 2013;138 222-234. DOI: 10.1016/j.biortech.2013.03.100.

[159] Moorthi PS, Selvam SP, Sasikalaveni A, Murugesan K, Kalaichelvan PT. Decolorization of textile dyes and their effluents using white rot fungi. African Journal of Biotechnology 2007;6(4) 424-429.

[160] Silveira E, Marques PP, Silva SS, Lima-Filho JL, Porto ALF, Tambourgi EB. Selection of Pseudomonas for industrial textile dyes decolourization. International Biodeterioration and Biodegradation 2009;63(2) 230-235. DOI: 10.1016/j.ibiod.2008.09.007.

[161] Daneshvar N, Ayazloo M, Khataee AR, Pourhassan M. Biological decolorization of dye solution containing Malachite Green by microalgae Cosmarium sp. Bioresource Technology 2007;98(6) 1176-1182. DOI: 10.1016/j.biortech.2006.05.025.

[162] Shah MP, Patel KA, Nair SS, Darji AM. Microbial decolourization of methyl orange dye by Pseudomonas spp. OA Biotechnology 2013;2(1) 10.

[163] Joe J, Kothari RK, Raval CM, Kothari CR, Akbari VG, Singh SP. Decolorization of textile dye Remazol Black B by Pseudomonas aeruginosa CR-25 isolated from the common effluent treatment plant. Journal of Bioremediation and Biodegradation 2011;2 118. DOI: 10.4172/2155-6199.1000118.

[164] Wu J, Jung BG, Kim KS, Lee YC, Sung NC. Isolation and characterization of Pseudomonas otitidis WL-13 and its capacity to decolorize triphenylmethane dyes. Journal of Environmental Sciences 2009;21(7) 960-964. DOI: 10.1016/S1001-0742(08)62368-2.

[165] Chen JP, Lin YS. Decolorization of azo dye by immobilized Pseudomonas luteola entrapped in alginate–silicate sol–gel beads. Process Biochemistry 2007;42(6) 934-942. DOI: 10.1016/j.procbio.2007.03.001.

[166] Işik M, Sponza DT. Effect of oxygen on decolorization of azo dyes by Escherichia coli and Pseudomonas sp. and fate of aromatic amines. Process Biochemistry 2003;38(8) 1183-1192. DOI: 10.1016/S0032-9592(02)00282-0.

[167] Chang JS, Chou C, Lin YC, Lin PJ, Ho JY, Hu TL. Kinetic characteristics of bacterial azo-dye decolorization by Pseudomonas luteola. Water Research 2001;35(12) 2841-2850. DOI: 10.1016/S0043-1354(00)00581-9.

[168] Voisard C, Keel C, Haas D, Dèfago G. Cyanide production by Pseudomonas fluorescens helps suppress black root rot of tobacco under gnotobiotic conditions. The EMBO Journal 1989 Feb;8(2) 351-358.

[169] Nandakumar R, Babu S, Viswanathan R, Raguchander T, Samiyappan R. Induction of systemic resistance in rice against sheath blight disease by Pseudomonas fluorescens. Soil Biology and Biochemistry 2001;33(4-5) 603-612. DOI: 10.1016/S0038-0717(00)00202-9.

[170] Samavat S, Heydari A, Zamanizadeh HR, Rezaee S, Aliabadi AA. Application of new bioformulations of Pseudomonas aureofaciens for biocontrol of cotton seedling damping-off. Journal of Plant Protection Research 2014;54(4) 334-339. DOI: 10.2478/jppr-2014-0050.

[171] Khan MR, Haque Z. Soil application of Pseudomonas fluorescens and Trichoderma harzianum reduces root-knot nematode, Meloidogyne incognita, on tobacco. Phytopathologia Mediterranea 2011;50(2) 257-266. DOI: 10.14601/Phytopathol_Mediterr-9252.

[172] Vanitha S, Ramjegathesh R. Bio control potential of Pseudomonas fluorescens against coleus root rot disease. Journal of Plant Pathology and Microbiology 2014;5(1) 1-4. DOI: 10.4172/2157-7471.1000216.

[173] Heil M, Bostock RM. Induced systemic resistance (ISR) against pathogens in the context of induced plant defences. Annals of Botany 2002:89(5) 503-512. DOI: 10.1093/aob/mcf076.

Chapter 6

Study of Cellular Processes in Higher Eukaryotes Using the Yeast *Schizosaccharomyces pombe* as a Model

Nora Hilda Rosas-Murrieta, Guadalupe Rojas-Sánchez, Sandra R. Reyes-Carmona, Rebeca D. Martínez-Contreras, Nancy Martínez-Montiel, Lourdes Millán-Pérez-Peña and Irma P. Herrera-Camacho

Additional information is available at the end of the chapter

http://dx.doi.org/10.5772/60720

Abstract

Schizosaccharomyces pombe (Sz. pombe), or fission yeast, is an ascomycete unicellular fungus that has been used as a model system for studying diverse biological processes of higher eukaryotic cells, such as the cell cycle and the maintenance of cell shape, apoptosis, and ageing. *Sz. pombe* is a rod-shaped cell that grows by apical extension; it divides along the long axis by medial fission and septation. The fission yeast has a doubling time of 2–4 hours, it is easy and inexpensive to grow in simple culture conditions, and can be maintained in the haploid or the diploid state. *Sz. pombe* can be genetically manipulated using methods such as mutagenesis or gene disruption by homologous recombination. Fission yeast was defined as a micro-mammal because it shares many molecular, genetic, and biochemical features with cells of higher eukaryotes in mRNA splicing, post-translational modifications as N-glycosylation protein, cell-cycle regulation, nutrient-sensing pathways as the target of rapamycin (TOR) network, cAMP-PKA pathway, and autophagy. This chapter uses *Sz. pombe* as a useful model for studying important cellular processes that support life such as autophagy, apoptosis, and the ageing process. Therefore, the molecular analysis of these processes in fission yeast has the potential to generate new knowledge that could be applied to higher eukaryotes.

Keywords: Yeast, *Schizosaccharomyces pombe*, cellular model, autophagy, apoptosis, ageing, chronological lifespan

1. Introduction

1.1. General features of *Schizosaccharomyces pombe*

Schizosaccharomyces pombe (*Sz. pombe*) is a yeast found in sugar-containing fermentations of alcohol involved in the production of beer in subtropical regions; the yeast was first described in 1893 by P. Lindner who isolated yeast from millet beer of East Africa and gave it the name "pombe" from the Swahili word for beer. *Sz. pombe* is a free-living fungus, classified with ascomyctes such as *S. cerevisiae* (Table 1). However, fission yeast is taxonomically and as evolutionarily distinct from budding yeasts as it is from humans. *Sz. pombe* reproduces by fission, a process similar to that used by higher eukaryotic cells. Despite the variation in the predicted time of divergence, phylogenetic analysis, and experimental evidence, it can be shown that the gene sequences of *Sz. pombe* are often more similar to its counterpart in mammals than the equivalent of genes of *S. cerevisiae* [1].

Sz. pombe is described as a fission yeast because it grows by apical extension and it divides along the long axis by medial fission and septation. The fission yeast has a doubling time of 2–4 hours, and can be maintained in the haploid or the diploid state. The yeast has a cylindrical rod-shape 3–4 μm in diameter and 7–14 μm in length [2]. The cell wall of *Sz. pombe* provides mechanical strength and protection from environmental stresses. The cell wall determines cellular morphology during the different stages of the life cycle, and is continuously remodelled during the cell cycle to allow cellular growth. The cell wall of *Sz. pombe* is composed of 74–82% glucan and 9–14% galactomannan, and has a chemical composition of β-1, 3-glucan, β-1, 6-glucan, α-1, 3-glucan, and α-galactomannan. Viewed by electron microscopy, the cell wall is organized in a three-layered structure: the outer layer is electron-dense, the adjacent layer is less dense, and the third layer bordering the cell membrane is dense [4, 5].

1.2. Genome organization of *Sz. pombe*

The whole genome of *Sz. pombe*, distributed over three chromosomes, is estimated at 13.8 Mb [6]. The genome of *Sz. pombe* was fully sequenced using 452 cosmids, 22 plasmids, 15 BAC clones, and 13 PCR products, in a project under the control of Schizosaccharomyces f European Sequencing Consortium (EUPOM), the Wellcome Trust Sanger Institute (United Kingdom), and Cold Spring Harbor Laboratory (USA). The collaborating bodies were Funding Agency Cancer Research UK (Cambridge Institute United Kingdom) and the European Union (EU) [7]. The genome is organized in the chromosomes I (5.7 Mb), II (4.6 Mb), and III (3.5 Mb), and a 20-kb mitochondrial genome. The three centromeres are 35, 65, and 110 kb long for chromosomes I, II, and III, respectively. On the centromeres are 174 tRNAs genes, 45 of which have introns. The genome of the yeast has at least 4, 940 protein-coding genes (including 11 mitochondrial genes) and 33 pseudogenes. *Sz. pombe* introns average only 81 nucleotides in length. The gene density for the complete genome is one gene every 2, 528 bp, which is similar in chromosomes II and I (one gene every 2, 483 and 2, 457 bp, respectively) and is less dense for chromosome III (one gene every 2, 790 bp). The spliceosomal RNAs (U1±U6) are located with 16 small nuclear RNA genes (snRNAs) and 33 small nucleolar RNAs (sno-RNAs). The 5.8S, 18S, and 26S ribosomal RNA genes are grouped as 100±120 tandem repeats in two arrays on chromosome III, with the 5S ribosomal RNA genes being distributed throughout the genome [7]. In silico analysis of the *Sz. pombe* genome has made it known that there are genes highly conserved between the vegetable and animal kingdoms, some of which have been implicated in cancers in humans [8, 9].

2. Mating-type locus, heterothallic, and homothallic phenotype

Sz. pombe has been studied extensively since the 1950s. Urs Leupold isolated the *Sz. pombe* strains widely used in research from a culture of *Sz. pombe* Lindner str. *liquefaciens* obtained by Osterwalder [10] from grape juice in France, in 1921. In its normal life cycle, *Sz. pombe* cells are haploids, although they can grow in a diploid state. Leupold described the haploid character and the complexity of the mating-type genetics of *Sz. pombe*, by culturing the haploid and diploid yeast [11, 12, 13], which were both homothallic and heterothallic (h) strains. The mating of *Sz. pombe* works by controlling *mat* genes located on the mating-type locus in the chromosome II [14], which is organized by one expression locus (*mat1*) and two donor loci *mat2* and *mat3* that encode the *mat P* and *mat M* genes, respectively. There are three alleles *mat* for the mating-type: h^+ (strain 975) or *P* (Plus, +), h^- (strain 972) or *M* (Minus, -), and h^{90} (strain 968). The mating-type is determined by the *mat1* locus, cells that present the segment called *mat1-P* determine the sexual type *P* or h^+ via the two genes *mat-Pi* and *mat-Pc* [15]. Cells with the segment *mat1-M* define the sexual type *M* or h^- by two genes *mat-Mi* and *mat-Mc*. The *P* and *M* alleles determine the mating reactions of the heterothallic strain. The homothallic strains with cells of both the *P* and *M* mating type can mate and form asci with four ascospores.

The heterothallic strain with only one mating type, are able to mate with heterothallic *Sz. pombe* strains of the opposite mating type. The allele h^{90} determine the homothallism [11], derived from *P* and *M* by recombination and mutation by the crossing of heterothallic strains. Homologous sequences flank the mating-type locus support site-specific recombination between the expression locus and the donor loci (*mat1* and *mat2*, or *mat1* and *mat3*) [16]. Heterothallic strains in *Sz. pombe* originated from strains having spontaneously rearranged their mating type *loci*. For homothallic segregants with the markers h^+ parent and recombination for two marker genes (*his7* and *his2*) flanking the locus, two mating-type genes are generated and named initially as *h1* and *h2* and later renamed *mat1* and *mat2*, respectively [17].

GENERAL FEATURES	
Doubling time of 2–4 hours	
Cylindrical rod-shaped	
3–4 μm in diameter and 7–14 μm in length	
Ascospores arranged linearly	
Easy genetic manipulation and mutant generation	
ORGANISM TAXONOMY	
NCBI Taxonomy ID: 284812	
Kingdom:	*Eukaryota*

Phylum:			*Ascomycota*	
Class:			*Schizosaccharomycetes*	
Order:			*Schizosaccharomycetales*	
Family:			*Schizosaccharomycetaceae*	
Genus:			*Schizosaccharomyces*	
Species:			*Schizosaccharomyces pombe*	
CELL CYCLE				
Phase			Percentage in the full cell cycle	
G1			10%	
S (*Cytokinesis*)				
G2			70%	
M				
MATING TYPES				
Fission yeast can exist stably in either haploid or diploid states.				
Phenotype			Symbol	
Homothallic			h^{90}	
Heterothallic P			h^{+}	
Heterothallic M			h^{-}	
GENOME				
Haploid Diploid		Gene density	Coding (%)	
	Length (Mb)			
Whole genome	13.8			
Chromosome	5.7	2483	58.6	Centromere I 35 kb
Chromosome	4.6	2457	57.5	Centromere II 65 kb
Chromosome	3.5	2790	54.5	Centromere III 110 kb
Mitochondrial genome	20 kb			
PEPTIDASES MEROPS				
Count of known and putative peptidases121 non-peptidase homologues 39				
PomBase (http://www.pombase.org) is the model organism database for the fission yeast *Schizosaccharomyces pombe*.				

Table 1. General features of *Schizosaccharomyces pombe*.

3. Cell cycle of *Sz. pombe*

Sz. pombe coordinate their growth with the cell cycle. Exponentially growing cells are of a similar cell size and double in mass before entering the next round of cell division. Cell length is a measure of the stage of the cell cycle that the cell has reached [18]. The cells are commonly cylindrical with rounded ends, but under conditions of starvation cells become shorter. *Sz.*

pombe is a single-celled, haploid organism with a cell cycle typical of a eukaryote, organized in the same stages as those of higher eukaryotes: interphase (G1, S, and G2 phases) and mitosis by mechanisms that are similar to those in multicellular animals. *Sz. pombe* usually proliferates in a haploid state. In a medium rich in nutrients, wild-type cells undergo a mitotic division approximately every two hours, with a short G1 phase taking less than 10% of the cell cycle. Next, the cells enters into the S phase following cytokinesis; the longer stage of the cell cycle is the G2 phase (70% of the cell cycle) [19]. Cells sense their size and use this information to coordinate cell division with cell growth to maintain a constant cell size by Pom1 protein and Nif1, localized to the cell ends of DNA, and inhibits mitotic activators in the middle of the cell acting as sensors of cell size controlling the onset of mitosis [20]. When starved of nitrogen, cells stop the division and arrest the cycle in G1, which promotes the sexual phase of the life cycle through conjugation of pairs of P and M cells to form diploid zygotes in a process known as karyogamy. In nitrogen-starved conditions, *Sz. pombe* haploid cells expressing the *mat P* and *mat M* genes enters into meiotic division after forming zygotes [21, 22]. After mating and nuclear fusion, premeiotic replication occurs, DNA content is duplicated from 2N to 4N, there is pairing and recombination of homologous chromosomes, and ocurrs the reductional meiosis I and the meiosis II division for the production of four haploid nucleic (1N), which are separated into four spores in an ascus, where the ascospores of *Sz. pombe* are arranged linearly [13]. *Sz. pombe* is normally haploid, but it is possible to select diploid strains, which can then divide by vegetative mitotic growth until starved of nitrogen, when they undergo meiosis and form azygotic asci. The provision of a rich nutrient source induces new vegetative growth and mitotic cell division. In 1957, Mitchison measured changes in the volume of cells along the yeast cell cycle. He found that volume increased slightly in the interphase, then changed about an hour before the cell split and increased just after cleavage. *Sz. pombe* extended at one end only, usually the older one [23, 24]. The cell cycle of *Sz. pombe* is described in Figure 1.

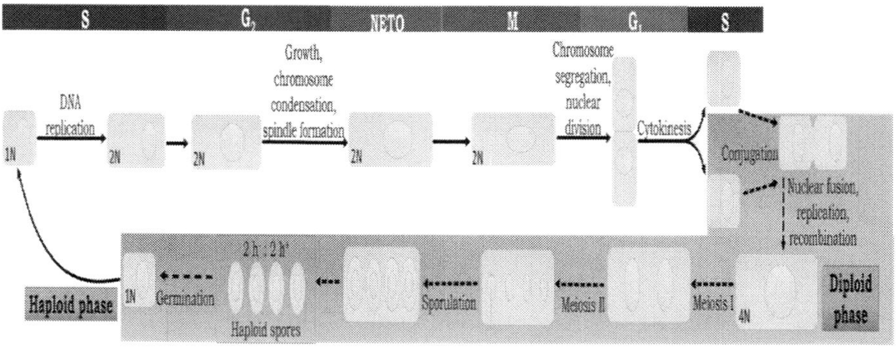

Figure 1. ***Sz. pombe cell cycle.*** DNA replication starts in S-phase then continues in G2-phase. *Sz. pombe* growth is accompanied by chromosome condensation and spindle formation. During G2, *Sz. pombe* starts bipolar growth known as New End Take Off (NETO), in M-phase chromosome segregation and nuclear division occurs, and in G1 cytokinesis is accomplished so two daughter cells are produced and it can undergo S-phase. When cells are starved, they undergo nuclear fusion, replication, and recombination, entering a diploid-phase which ends in the formation of four haploid spores that, under normal conditions, enter into S-phase.

4. Cellular process studied using *Sz. pombe*

Sz. pombe is an excellent model for the study of common eukaryotic cellular processes. The yeast shares important biological aspects with metazoans, such as chromosomal structure and metabolism, relatively large chromosomes and centromeres, low-complexity replication origins, epigenetic mechanisms for regulation of genetic expression, G2/M cell cycle control, cytokinesis, mitosis and meiosis, DNA repair and recombination, the mitochondrial translation code, and spliceosome components, among others [2, 25].

Fission yeast presents a short cell cycle and is easy to manipulate by genetic classical and molecular analysis. *Sz. pombe* is a particularly useful model for the study of the function and regulation of genes from more complex species [26]. A model organism is any species used for scientific research in order to answer a specific question that cannot be studied in any other way. A model must be amenable to experimental manipulation in the laboratory, require low costs of culture and maintenance, have a short generation time, be amenable to genetic manipulation to produce mutants and allow the study of their effects, and show high conservation of mechanisms or specific cellular process. A system model extensively used in microbiology, biochemistry, biotechnology, and molecular biology is *Escherichia coli* (*E. coli*). Although this bacterium can be grown and be cultured easily and inexpensively in a laboratory setting, *E. coli* lacks the mechanisms for post-transcriptional on mRNA or post-translational modifications on proteins of eukaryotic origin. *Saccharomyces cerevisiae* (*S. cerevisiae*) is other model used due to its quick, inexpensive, and easy growth in control conditions of the laboratory. However, like *E. coli*, *S. cerevisiae* has a limited capability for performing post-transcriptional or post-translational modifications such as N-glycosylations. *Drosophila melanogaster* is other model system in the cellular biology field widely used for classical genetics analysis. *Caenorhabditis elegans* is the model of choice for understanding the genetic control of development, physiology, and the study of cellular process such as apoptosis and ageing. *Mus musculus* and *Rattus norvegicus* are two mammalian models used for the study of several diseases that affect humans; produced by highly conserved genes, its main disadvantage is slower reproduction than other models allow. In contrast, fission yeast is an ideal system for the analysis of cellular processes common in eukaryotic cells; since it shares many molecular, genetic, and biochemical characteristics with multicellular organisms — such as the splicing of mRNA, post-translational modifications, cell division, and cell cycle control [13, 27, 28] — it has a high rate of reproduction in a short period.

4.1. Model for studying the eukaryotic cell cycle

Sz. pombe, the fission yeast, has been used as a model for the study of the eukaryote's cell cycle. Fission yeast is a powerful tool for the study of the cell cycle, since it has a typical cycle cell eucarionte G1, S, G2, and M. G2 occupies approximately 70% and the remaining stages on 30% of the total duration of the cell cycle [29, 30]. Since it has a typical cell shape, it is a useful system for identifying genes involved in the cell cycle because mutants are easily identifiable when compared with the wild type [31]. In the 1970s, using *Sz. pombe* P. Nurse demonstrated the cyclin/cyclin-dependent kinase complex as the key regulator of cellular cycle. *Sz. pombe* has a single mitotic cyclin-dependent kinase (CDK), highly conserved, named cdc2p, homologous to CDC28 from *S. cerevisiae* and CDK1 from human cells [32]. *Sz. pombe* has been used for the molecular sexual differentiation analysis. In this system, it was discovered that a cluster of genes associated with the main biological events of the differentiation process are up-regulate: genes induced in response to environmental changes (starvation and pheromone-induced genes), early genes (pre-meiotic S phase and recombination), middle genes (meiotic divisions and early steps of spore formation), and late genes (spore maturation) [33]. In addition, the use of the fission yeast has contributed to the knowledge of other areas such as cytokinesis, the formation of microtubules, differentiation, cell morphogenesis and polarity, mechanisms of response to stress, and response to damage of the DNA [34, 35].

4.2. Cellular ageing

The identification of evolutionarily conserved mechanisms that determine long lifespans has been of great interest, making the fission yeast *Sz. pombe* a novel model for the study of ageing. Ageing is a genetically programmed event [36, 37] and can be defined as a decline in biological functions due to the accumulation of molecular and cellular damage imposed by intrinsic and extrinsic factors on living organisms, having mortality as a consequence. Lifespans of humans are measured chronologically from birth to death. Ageing can considered from the perspectives of two types of cellular lifespans: the replicative lifespans (RLS) of continuously dividing cells and the chronological lifespans (CLS) of post-mitotic cells, from terminally differentiated cells. RLS is the number of mitotic divisions a cell can undergo before senescence, and CLS is the length of time a non-dividing cell can remain viable. CLS measures how long cells can remain viable in a non-dividing state, allowing a continuous decline in viability without detectable re-growth until all cells in the culture are dead. The complexity of human bodies and the length of the human lifespan make research on the biology of human ageing difficult. Model organisms such as yeast, worms, and flies have been extensively utilized because of their short lifespans and their accessibility to classical genetics techniques, as well as the high degree of conservation of genes, processes, and signalling pathways that they share with humans and other mammals [38]. Some aspects of ageing studied in fission yeast have been mainly focused on the effect of calorie restriction on ageing and on the detection of nutrient signalling pathways [39, 40]. The protein kinase target of rapamycin (TOR) and insulin/insulin-like growth factor (IGF)-1 pathway are the well-studied nutrient sensing pathways that determine the length of lifespan [41]. TOR exerts its function through two distinct complexes: TOR complex 1 (TORC1) and TOR complex 2 (TORC2), which regulate translation, cell growth, and the organization of actin cytoskeleton [42, 43]. Inactivation of TORC1 increases the lifespan in model organisms such as budding yeast, worms, flies, and mice [44, 45]. The insulin/IGF-1 signalling pathway, and their components such as PI3 kinase and AKT kinase, affect the longevity of flies, worms, and mice [39]; interestingly, this pathway has also been found in yeasts with the AKT kinase. In *S. cerevisiae*, nutrient signalling via the kinases Sch9/Tor/PKA pathway has a strong pro-ageing effect. Down-regulation of this pathway is partially responsible for CLS increase. In fission yeast, there are two homologues of these kinases with a similar pro-ageing activity in *Sz. pombe*: pka1 y sck2 [46]. In this model, deletion of *pka1* or *sck2* extended CLS by diminishing the accumulation of ROS and increasing oxidative stress resistance [47,

48]. AKT kinase orthologues of *Sz. pombe* regulate CLS: loss of *sck1* extended the lifespan in the condition with the excess of nutrients and loss of both genes showed that *sck1* and *sck2* control different longevity pathways. In addition, there are four genes identified as regulators of ageing in *Sz. pombe*: 1. *psp1+/sds23+/moc1+*, a gene required for long-term survival in the stationary phase. Psp1 is phosphorylated in the stationary phase by the cyclin-dependent kinase complex Cdc2/Cdc13. 2. Casein kinase *cka1+*, an orthologue of the α-subunit of casein kinase II in mammals, whose function is essential but the details of the effect of its deletion are not known. 3. *adh1+*, encoding the alcohol dehydrogenase that converts acetaldehyde to ethanol. 4. *rpb10+*, a protein subunit shared by RNA polymerases I, II, and III. It is noteworthy that lifespan extension from TOR-inhibition, Sch9/Akt, or caloric restriction is an effect highly conserved between yeast, worms, and flies [39]. Research of ageing on fission yeast has greatly benefited the knowledge of molecular process since the data obtained can be extrapolated to mammalian cells.

4.3. Autophagy

Autophagy is a catabolic mechanism that regulates the intracellular turnover of unfolded/misfolded, long-lived, or damaged proteins, lipids, and organelles (such as mitochondria and the endoplasmic reticulum (ER), through its sequestration within a double-membrane and delivery to lysosomes for degradation and recycling of biocomponents. Under basal conditions, autophagy is a housekeeping programme, but it can also be activated by nutrient starvation, low cellular energy levels, amino acid deprivation, growth factor withdrawal, ER stress, hypoxia, oxidative stress, and infection [50–54]. Autophagy deregulation in higher eukaryotes leads to muscle atrophy, myopathy, and cardiac and immune disease [52–54], and plays a dual role in the pathogenesis of cancer [50, 52, 55, 56].

Duve [57] introduced the term autophagy from the Greek *auto* (itself) and *phagy* (eat) which means that cells degrade their cytoplasmic components through lysosomal degradation. Clark described the autophagy process in 1957 through electron microscopy of the kidneys of newborn mice [58]; he discovered that inside lysosomes there were mitochondria. Unfortunately, it was very difficult to study autophagy via electron microscopy and biochemical analysis; thanks to the introduction of molecular techniques, it was later discovered that the yeast *S. cerevisiae* was a tractable model of eukaryotes. Thirty Atg proteins (autophagy-related proteins) were also discovered later, and it was found that they possess mammalian orthologues [50–52, 54, 59–61]. Four types of autophagy have been described in mammals (Figure 2): macroautophagy (hereafter referred to as autophagy), microautophagy, chaperone-mediated autophagy (CMA), and the more recently described RNautophagy [61–64]. In autophagy, the cytoplasmic components (also known as cargo) are first encapsulated in a double-membrane structure synthesized *de novo*, which at the beginning is the phagophore and when it is completely formed is called autophagosome (AP), after which it is trafficked and fused to the lysosome. In microautophagy the cargo is selected through an invagination in the lysosome where it is degraded. In the CMA, specific cytosolic proteins containing the motif KFERQ are recognized by the heat shock cognate protein 70 (HSC70), and are then delivered to the lysosomal associated membrane protein (LAMP2) of type A receptor that is located in the lysosome membrane, where the proteins are translocated inside and degraded [62–64]. In 2013, Fijiwara [65] described RNautophagy, where several RNAs were recognized by the LAMP2C receptor and degraded in the lysosome. However, autophagy is a very complex process in metazoans, making it important to have simplest biomodels that share with mammals the core of the autophagic machinery, such as the budding yeast *S. cerevisiae* and the fission yeast. In the budding yeast autophagy, microautophagy and cytoplasm to vacuole targeting (CVT) have been outlined; likewise in the fission yeast, autophagy shares almost the same conserved core of autophagic machinery (Figure 2). However, microautophagy, CVT, and RNautophagy have not been described and further investigation is needed in order to unravel the conservation of these mechanisms.

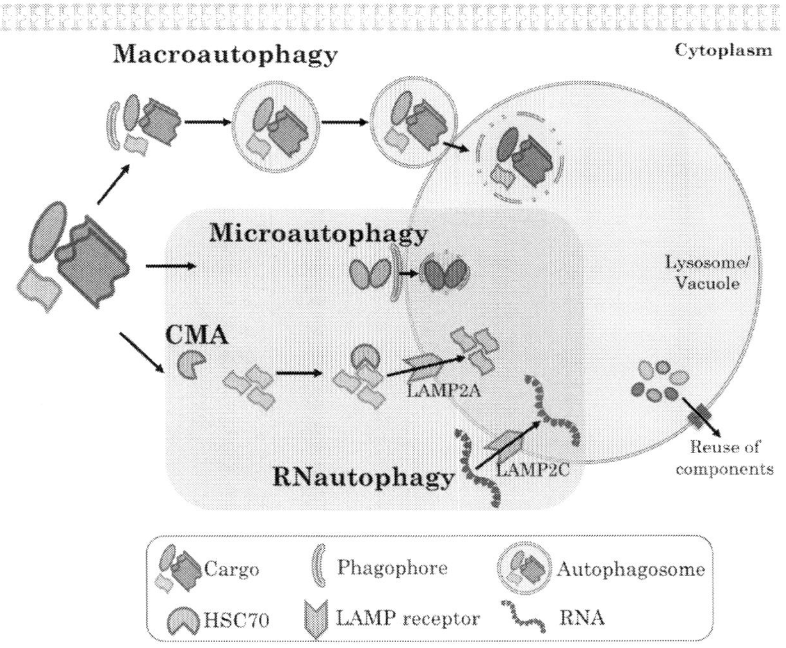

Figure 2. Types of autophagy. In mammals (*H. sapiens, M. musculus, and R. norvergicus*), macroautophagy, microautophagy, cytoplasm to vacuole targeting (CVT), and RNautophagy have been described. In macroautophagy, cargo is sequestered by an invagination of a lipid bilayer which at the beginning is called phagophore, before it elongates and madurates, forming the autophagosome, then it fuses to the lysosome. In microautophagy, cargo is selected by an invagination of the lysosome. In CVT, proteins with the motif KFERQ are recognized by the HSC70, delivered, and translocated inside the lysosome. In RNautophagy (in *Sz. pombe*), macroautophagy is highly conserved; however, microautophagy, CVT, and RNautophagy (in the blue square) have not yet been elucidated.

The Atg machinery was classified into five groups according to function in each step of the autophagic process (Table 2 and Figure 3) [1–6, 11, 12, 50–55, 60, 63, 64]: ULK1 kinase and its regulators that signal the autophagosome biogenesis, the phosphatidylinositol (PtdIns) 3-kinase complex controls the nucleation step that recruitsother Atg proteins hierarchically, the elongation of the phagophore mediated by the Atg12 and LC3 conjugation systems, and a subgroup of proteins with unknown functions. In mammals under nutrient rich conditions, insulin or growth factors activate the phosphatidylinositol 3-kinase (PI3K)/Akt/mammalian target of rapamycin (mTOR) pathway. When PI3K is activated, it converts phosphatidylinositol (3, 4)-bis phosphate (PIP2) to phosphatidylinositol-3 phosphate (PIP3), then Akt is activated and inactivated by phosphorylation to TSC which leads to mTOR complex 1 (mTORC1) activation. mTORC1 negatively regulates ULK1 and Atg13, so when mTORC1 is inactivated by nutrient depletion, low concentration of insulin, deprivation of amino acids, or the addition of its antagonist rapamycin, the ULK1 kinase and Atg13 are then activated, forming the ULK1 kinase complex (Table 2), which signals the induction of the AP formation. There is another inductor of autophagy, the AMP activated protein kinase (AMPK), that senses low intracellular energy status and directly activates by phosphorylation to ULK1, and by inhibiting mTORC1 via phosphorylation of Raptor (a subunit of mTORC1) [66–69]. ULK1 phosphorylates Beclin1 at serine 14, which is a component of the PtdIns 3- kinase complex (Table 2) and enhances Vps34 activity — this step is crucial in the AP formation because Vps34 inhibition by 3-methyladenine (3-MA) or wortmannin disrupts the biogenesis. Vps34 synthesizes PIP3 at the sites where AP are assembled; it has been suggested that PIP3 recruits additional factors for AP formation, but its role remains unclear. Beclin1 binding proteins that activate or inhibit AP biogenesis have been identified: the UV radiation resistance-associated gene protein (UVRAG) and Bcl-2 or Bcl-X_L, respectively. Beclin-UVRAG interacts with the class C Vps complex proteins, which are part of the endosomal fusion machinery, so this interaction induces AP fusion with lysosomes [50, 53, 70–71].

The next step in AP building is the elongation of the phagophore, which requires membrane input from organelles (such as ER, mitochondria, cytoplasmic membrane, or possibly from *de novo* synthesis); however, the exact membrane source is unknown. In mammals, Atg9 is a transmembrane protein that probably carries membrane expansion of the phagophore but its downstream effectors have not yet been elucidated. However, there are two ubiquitin-like complexes that are required for elongation; the first one is the Atg12 conjugation system (Table 2), in which the ubiquitin-like protein Atg12 is covalently joined to Atg5. This must first be activated by Atg7 (E1 ubiquitin activating enzyme-like or E1). Atg7 uncovers Atg12 glycine carboxyl-terminus, then Atg12 is transferred to Atg10 (E2 ubiquitin conjugation enzyme-like or E2), after which it is covalently attached by its terminal glycine to lysine 130 (Lys 130) of Atg5.

Finally, Atg12-Atg5 conjugates with Atg16L1 forming a tetramer; this tetramer is essential for the elongation but when AP has been completed it dissociates [76–78]. The second ubiquitin-like complex is the microtubule-associated protein 1 light chain 3 (LC3) conjugation system (Table 2). LC3 is a precursor of LC3-I, it is obtained by the cleavage of the protease Atg4B to its carboxyl terminus. LC3-I is attached to phosphatidylethanolamine (PE) by the enzymes E1 and E2- ubiquitin-like Atg7 and Atg3, respectively, obtaining the LC3-II product which forms part of the AP. Notice that the LC3 complex lacks an E3 ubiquitin ligase-like enzyme that could facilitate LC3I-PE conjugation; however, a crosstalk between these two ubiquitin-like systems has been demonstrated by the tetramer Atg12-Atg5-Atg16L1 acting as E3-like enzyme. When AP is formed, LC3 can be recycled from the form LC3II; this is achieved through Atg4 which breaks apart LC3I from PE, and LC3-I and LC3-II assays are widely used for monitoring autophagy [60, 77]. Finally AP moves bidirectionally along microtubules towards the microtubule organizing centre (MTOC) where lysosomes are enriched, then AP fuses with lysosomes forming autolysosomes where the content is degraded and the components are recycled [50]. Atg2 is required for the AP location and can possess a different role when binding to Atg18 in order to target the complex to autophagic membranes [79]. In *Sz. pombe,* depletion of nitrogen from the culture medium triggers autophagy. Unlike mammals, the addition of AMPK does not induce autophagy, nor can rapamycin induce it. It has been reported that rapamycin has a partially inhibitory effect on TOR activity, moreover, the addition of rapamycin plus caffeine prompts a starvation-specific gene expression pattern and autophagy [80–81]. Atg 1, 2, 3, 4, 5, 6, 7, 8, 9, 11, 12, 13, 14, 15, 16, 18a, 18b, 18c, Vps34 homologues have been found (Table 2) [73–75], and their deletion inhibits AP formation.

	Components		Features
	Mammals % identity	*Sz. pombe* (% identity)	
ULK1 kinase complex	ULK1 *M. musculus*: 89% *R. norvergicus*: 89%	Atg1 38%	Kinase ser/thr induces autophagy and PA assembly [72, 73–75]
	Atg13 *M. musculus*: 94% *R. norvergicus*: 89%	Atg13 23%	Regulator and binding protein of ULK1 complex [72, 73–75]
	FIP200 *M. musculus*: 89% R. norvergicus: 89%	Atg17 47%	Binding protein and regulator of ULK1, induces PA biogenesis [72, 73–75]
	Atg101 *M. musculus*: 99% *R. norvergicus*: 98%	Atg101 25%	Binds to Atg13 in the ULK-Atg13-FIP200 complex, which is important for the stability and basal phosphorylation of Atg13 and ULK1 [72, 73–75]
PtdIns 3- kinase complex	Vps15 *M. musculus*: 96% *R. norvergicus*: 96%	Vps15 44%	Regulatory kinase subunit of the PtdIns 3- kinase complex [72, 73-75]
	Atg14 *M. musculus*: 93% *R. norvergicus*: 98%	Atg14 32%	Necessary for the function of the PtdIn2-kinase complex [72, 73–75]
	Vps34 *M. musculus*: 97% *R. norvergicus*: 98%	Vps34 52%	PtdIns 3- kinase produces PI3P and allows the recruitment of PI3P binding proteins like WIPI1/2 and the two conjugation systems [72, 73–75]
	Beclin1 *M. musculus*: 98%	Atg6 29%	Modulates AP biogenesis by binding to Vps34 [72, 73–75]

	Components		Features
	R. norvergicus: 97%		
	UVRAG *M. musculus*: 88% *R. norvergicus*: 94%	**Vps38** 25%	Binds Beclin1 enhancing AP biogenesis [72, 73–75]
Atg12 conjugation system	**Atg12** *M. musculus*: 76% *R. norvergicus*: 71%	**Atg12**	Ubiquitin-like protein covalently joined to Atg5 [72, 73–75]
	Atg5 *M. musculus*: 97% *R. norvergicus*: 95%	**Atg5** 28%	Ubiquitin-like protein covalently attached to Atg12 and interacts with Atg16 [72, 73–75]
	Atg7 *M. musculus*: 89% *R. norvergicus*: 89%	**Atg7** 38%	E1-like enzyme for Atg12-Atg5 and LC3-PE formation [72, 73–75]
	Atg10 *M. musculus*: 72% *R. norvergicus*: 79%	**SpAtg10** 21%	E2-like enzyme required for Atg12-Atg5 complex In *Sz. pombe,* SpAtg10is also required for normal cell cycle progression and for responses to several stress agents but has no role in autophagy [72, 73–76]
	Atg16 *M. musculus*: 82% *R. norvergicus*: 82%	**Atg16** 0%	Necessary for the assembly of the tetramer Atg12-Atg5-Atg16 [72, 73–75]
LC3 conjugation system	**LC3** *M. musculus*: 83% *R. norvergicus*: 83%	**Atg8** 37%	Conjugated with PE forms part of the AP membrane [72, 73–75] [22, 23–25]
	Atg3 *M. musculus*: 91% *R. norvergicus*: 93%	**Atg3** 36%	E2-like enzyme [72, 73–75] [22, 23–25]
	Atg4 *M. musculus*: 92% *R. norvergicus*: 91%	**Atg4** 28%	Induces LC3-II then helps to recycle LC3 [72, 73–75]
	Atg7 *M. musculus*: 89% *R. norvergicus*: 98%	**Atg7** 38%	E1- like enzyme [72, 73–75]
Atg2-Atg18 complex and Atg9	**Atg2** *M. musculus*: 85% *R. norvergicus*: 84%	**Atg2A** 27%	Required for AP localization [72, 73–75]
		Atg2B	Required for AP localization [72, 73–75]
	WIPI *M. musculus*: 94% *R. norvergicus*: 94%	**SpAtg18a** 34%	In *Sz. pombe* is a protein required for the targeting of Atg12-Atg5-Atg16 complex [72, 73–75]
		SpAtg18b	

Components		Features
	29%	
	SpAtg18c 25%	
Atg9 *M. musculus*: 98% *R. norvergicus*: 98%	Atg9 29%	Transmembrane protein that might carry membrane expansion of phagophore [72, 73–75]

*% of identity respect to Atg *Homo sapiens* proteins, obtained using Blast http://blast.ncbi.nlm.nih.gov/Blast.cgi

Table 2. Autophagic machinery in mammals and *Sz. pombe.*

Finally, Atg12-Atg5 conjugates with Atg16L1 forming a tetramer; this tetramer is essential for the elongation but when AP has been completed it dissociates [76–78]. The second ubiquitin-like complex is the microtubule-associated protein 1 light chain 3 (LC3) conjugation system (Table 2). LC3 is a precursor of LC3-I, it is obtained by the cleavage of the protease Atg4B to its carboxyl terminus. LC3-I is attached to phosphatidylethanolamine (PE) by the enzymes E1 and E2- ubiquitin-like Atg7 and Atg3, respectively, obtaining the LC3-II product which forms part of the AP. Notice that the LC3 complex lacks an E3 ubiquitin ligase-like enzyme that could facilitate LC3I-PE conjugation; however, a crosstalk between these two ubiquitin-like systems has been demonstrated by the tetramer Atg12-Atg5-Atg16L1 acting as E3-like enzyme. When AP is formed, LC3 can be recycled from the form LC3II; this is achieved through Atg4 which breaks apart LC3I from PE, and LC3-I and LC3-II assays are widely used for monitoring autophagy [60, 77].

AP moves bidirectionally along microtubules towards the microtubule organizing centre (MTOC) where lysosomes are enriched, then AP fuses with lysosomes forming autolysosomes where the content is degraded and the components are recycled [50]. Atg2 is required for the AP location and can possess a different role when binding to Atg18 in order to target the complex to autophagic membranes [79]. In *Sz. pombe*, depletion of nitrogen from the culture medium triggers autophagy. Unlike mammals, the addition of AMPK does not induce autophagy, nor can rapamycin induce it. It has been reported that rapamycin has a partially inhibitory effect on TOR activity, moreover, the addition of rapamycin plus caffeine prompts a starvation-specific gene expression pattern and autophagy [80–81]. Atg 1, 2, 3, 4, 5, 6, 7, 8, 9, 11, 12, 13, 14, 15, 16, 18a, 18b, 18c, Vps34 homologues have been found (Table 2) [73–75], and their deletion inhibits AP formation.

4.4. Apoptosis

The maintenance of homeostasis in pluricellular and unicellular organisms is achieved through lots of mechanisms, one of the most important of which is related to the death of the cell itself. In this way, cell death regulates the number of cells in a tissue or a colony, and the removal of damaged cells. In 1963, Lockshin introduced the term Programmed Cell Death (PCD) in order

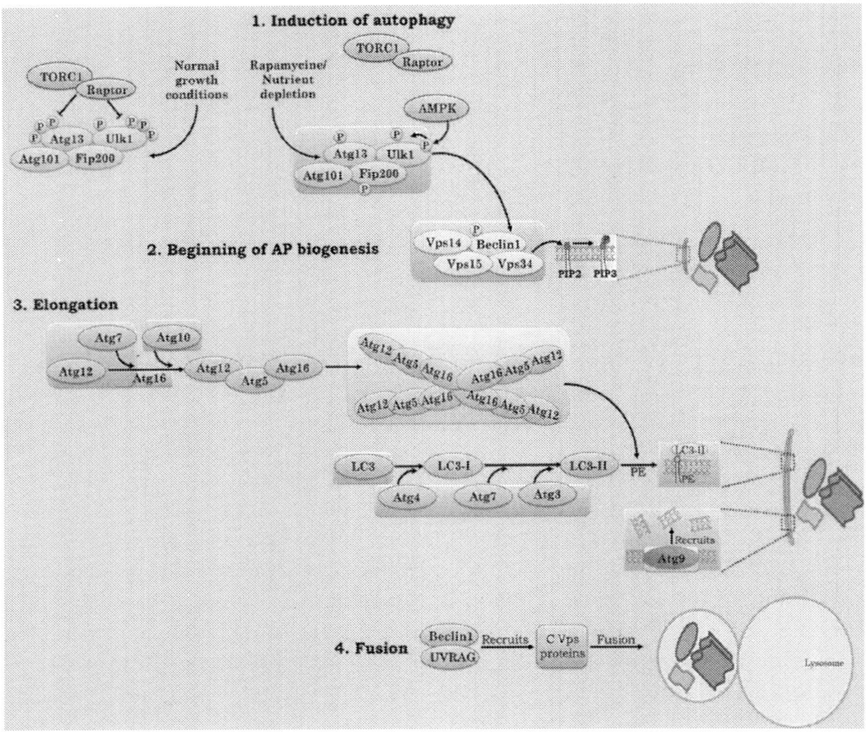

Figure 3. Mechanism of the core Atg proteins in autophagy. Atg proteins with a purple rectangular background have been identified in *Sz. pombe*, the ones in blue rectangles have not yet been described, and the ones in a red rectangle have been identified but have no role in autophagy.

to define the programmed and controlled self-destruction process in a local and temporal way [83–84]. There are many ways for a cell to die; one of them is via apoptosis. The term apoptosis was introduced by Kerr and colleagues in 1972, which is defined as a highly coordinated cellular suicide programme controlled principally by zymogens (i.e., caspases and metacaspases), proteins of the Bcl-2 family, and mitochondria [85]. The Nomenclature Committee on Cell Death established the morphological and biochemical changes in the apoptotic cell, such as membrane blebbing, cell shrinkage (pyknosis), chromatin condensation plus fragmentation (karyorrhexis), formation of membrane bound cell fragments (apoptotic bodies), decrease in mitochondrial inner transmembrane potential, selective cleavage of various cellular proteins, and the translocation of phosphatidylserine from the inner to the outer leaflet of the plasma membrane [86].

The importance of Bcl-2 proteins relies on Cyt C, which in basal conditions works at the respiratory chain but plays a different role in the cytoplasmic space: Cyt C oligomerizes with the adaptor protein Apoptosis protease activating factor-1 (Apaf-1), forming the complex known as apoptosome [98, 99]. The apoptosome activates by proteolytic cleavage to caspase-9 which is one of the cascade cysteine-aspartate proteases (caspases); in other words, caspases are liberated from their inhibitory prodomain [98, 99]. Once caspase-9 is activated, the downstream zymogens procaspase 3/7 are activated as caspase 3/7, then they attach to specific substrates in the cell leading to cell dismantling but they must first be released from their endogenous inhibitor X-linked inhibitor of apoptosis (XIAP). XIAP activity is inhibited by SMAC/ DIABLO and HtrA2/Omi, which should be remembered as mitochondrial proteins that were released along Cyt C [94, 95, 100]. In *C. elegans* the mitochondrial permeabilization and the release of Cyt C have no role in CED-3 activation, unlike its mammalian counterparts. Another difference between the nematode and mammals is that CED-4 lacks a Cyt C interaction motif, this means that some of the apoptotic machinery are conserved among them. However, some protein-protein interactions differ from each other.

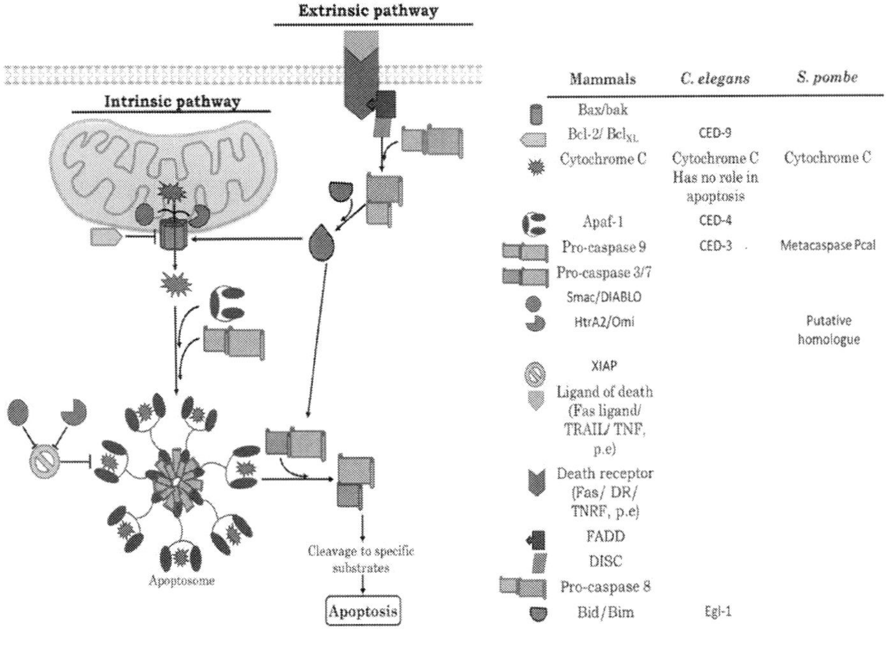

Figure 4. Apoptosis in eukaryotic cells. The factors involved in the basic molecular machinery executing cell death are conserved in yeast, including the yeast caspase YCA1. Apoptosis in yeast is composed of molecules shared with other model systems. This figure shows the intrinsic and extrinsic pathways as well as the components from the *C. elegans*, mammalians, and *Sz. pombe*.

On the other hand, the extrinsic pathway start when death ligands such as tumour necrosis factor (TNF), Fas ligand (FasL), TNF-related apoptosis-inducing ligand (TRAIL), among others, join and activate their respective transmembrane death receptor, such as Fas. The interaction of death ligands with death receptors result in the recruitment of adaptor proteins like the Fas-associated death domain protein (FADD), which recruits, aggregates, and promotes procaspase-8 activation. Caspase-8 switch on procaspase 3/7 and the Bid protein through is proteolysis. At this point, a crosstalk between the extrinsic and intrinsic pathways are mediated by Bid activation since truncated Bid (tBid) promotes Cyt-C by interacting with Bax, leading to Bax insertion in the mitochondrial membrane as an oligomer pore [93–95].

Study of apoptosis is difficult in higher eukaryotes due to the complexity of the phenomena itself, which is why it was thought until recently that unicellular organisms could not have PCD machinery because it would mean that the organism could orchestrate its suicide, so yeast were employed as purely naïve backgrounds for studying proteins' interaction. In 1997, B. Ink and JM. Jurgensmeier independently performed a yeast two-hybrid system in *Sz. pombe* to study the proteins of the apoptotic cascade. It was found that the expression of the proteins Bax and Bak lead to the induction of chromatin condensation and DNA cleavage [101–102]. Moreover, the lethal phenotype was obliterated by the co-expression of the antiapoptic Bcl-$_{XL}$ and Bcl-2, therefore, yeasts undergo mammalian-like apoptosis. Likewise, C. James proved that CED-4 expression in the fission yeast induced chromatin condensation and it was abolished by CED-9 interaction [103]. Induction of a yeast apoptotic programme can be triggered by exogenous and intrinsic stresses, such as low doses of H_2O_2, acetic acid, valproic acid, HCl, and NaOH in the media, cell ageing, and temperature stress [104–107]; these stimuli induce ROS that also enable an apoptosis programme. Additionally, DNA damage-responsive proteins are highly conserved among fission yeast and metazoans; hence, defects in DNA replication promote high levels of ROS, promoting apoptosis. Interestingly, SpRad9 — the orthologue of the human Rad9 — controls cell cycle checkpoints, and also possess a BH3 domain which enables binding with Bcl-2 and Bcl-XL in mammal cells, prompting cell death. This proves that the Bcl-2 family of proteins may be evolutionarily conserved from fission yeast to mammals [108–112].

Caspases are the main components of the apoptotic pathway, and are not found in *Sz. Pombe*; however, it possess an orthologue in the metacaspase PcaI (Figure 4). Unlike mammalian caspase activation, PcaI undergoes auto proteolytic cleavage activation just as in metazoan caspases; however, until now there has been no evidence of any substrate of PcaI [104, 105, 113]. Notwithstanding, overexpressing PcaI in *Sz. pombe* does not induce an apparent cell death phenotype, moreover it appears to stimulate growth. Also, PcaI expression was up-regulated by oxidative and heavy metal toxicity, heat, hyperosmotic stress and DNA damage conditions, H_2O_2, sodium nitroprusside, and CdCl, giving cytoprotection; this evidence suggests that the metacaspase could be involved in growth and stress adaptation of *Sz. pombe*. It has also been demonstrated that the level of oxidized proteins is proportional to the presence of the metacaspase, possibly due to a cleaning function within the population. Apparently, PcaI has pro-survival and pro-apoptotic roles and further investigation is needed to understand its function [105, 113, 114]. The Htra2/Omi-like protein Nma11p (Nuclear mediator of apoptosis) is found in *S. cerevisiae*, which is a serine protease, having a pro-apoptotic function, and which aggregates in the nucleus after cellular stress conditions; in *Sz. Pombe*, its putative homologue has been found [115]. In mammals, an apoptotic marker is the release of CytC; as in yeasts, its presence in the cytoplasm is found after the induction of the apoptotic programme. Just as in mammals, the release of CytC decreases mitochondrial inner transmembrane potential, but unlike metazoans the upstream and downstream effectors of PcaI have not been elucidated. The release of CytC has been correlated with the induction of cell death in *S. cerevisiae* [116], which correlates highly with the mammalian intrinsic pathway; nevertheless, CytC release has yet to be investigated in *Sz. pombe*.

5. *Sz. pombe*: Heterologous expression systems of proteins

Fission yeast is a useful system for studying the function and regulation of genes, and an excellent host for heterologous expression of molecules derived particularly from higher eukaryotes because it produces high cell densities, short fermentation times, and the use of chemically defined media [117]. Difficulties in the production and purification of heterologous proteins are related mainly to proteolytic degradation of gene products by specific proteases of the host [118]. This makes it highly important to obtain strains that have specific mutations eliminating the activity of proteases. Success of this expression is based on the use of genome-reduced strains with the deletion of unnecessary genes. Gene-deletion technology used for this purpose is the LATOUR (latency to universal rescue) method, which eliminates a DNA segment of more than 100 kb [119, 120]. This strategy of gene deletion is useful for identifying genes that are essential to the yeast [121]. The *Sz. pombe* proteolytic system has been seldom studied. Currently, two proteolytic systems have been identified with different subcellular localizations: a vacuolar one (where can be found the Proteinase yspA, Proteinase yspB, Carboxipeptidasa Y1p, Aminopeptidase yspI, Dipeptidyl aminopeptidase DPK, and DPC) and another Cytosolic level (Leucin Aminopeptidase yspII, peptidase multicorn, and proteasome 26S) [122, 123]. From full-genome sequence data which are deposited in the MEROPS database, an output of 121 peptidases and probable peptidases, being non-peptidase homologues 39 proteases, has been identified [124].

6. Conclusions

This chapter discussed some topics related to the employment of *Sz. pombe* as a model system for the study of the cellular processes of eukaryotes. Despite the complexity of mammalian cells with respect to single-celled yeasts, many intracellular processes are highly conserved, such as cell cycle, autophagy, and ageing; there are even similarities at the level of the mechanisms of the processes carried out in the cell. In some cases, there is only the sharing of some genes, as in the apoptosis process. However, their genetic regulation could be similar in triggering the specific cellular responses. The use of yeast *Sz. pombe* as a model of study allows the generation of new knowledge with direct application to mammalian cells such as human ones.

Author details

Nora Hilda Rosas-Murrieta[1*], Guadalupe Rojas-Sánchez[1], Sandra R. Reyes-Carmona[1], Rebeca D. Martínez-Contreras[2], Nancy Martínez-Montiel[2], Lourdes Millán-Pérez-Peña[1] and Irma P. Herrera-Camacho[1]

*Address all correspondence to: nora.rosas@correo.buap.mx

1 Laboratorio de Bioquímica y Biología Molecular, Centro de Química, Instituto de Ciencias, Benemérita Universidad Autónoma de Puebla. Edif. 103F, Ciudad Universitaria-BUAP, Jardines de San Manuel, Puebla, México

2 Centro de Investigación en Ciencias Microbiológicas, Instituto de Ciencias, Benemérita Universidad Autónoma de Puebla, México

References

[1] Rhind N, Chen Z, Yassour M, et al. (2011) Comparative functional genomics of the fission yeasts. *Science*. 332(6032): 930–936.

[2] Forsburg SL. (2005) The yeasts *Saccharomyces cerevisiae* and *Schizosaccharomyces pombe*: models for cell biology research. *Gravit Space Biol.* 18(2): 3–10.

[3] Moreno S, Klar A, Nurse P. (1991). Molecular genetic analysis of fission yeast Schizosaccharomyces pombe. *Methods in Enzymol*. 194: 795–823.

[4] Bush DA, Horisberger M, Horman I, et al. (1974). The wall structure of Schizosaccharomyces pombe. *Journal of General Microbiology*. 81: 199–206.

[5] Humbel BM, Konomi M, Takagi T, et al.(2001) In situ localization of beta-glucans in the cell wall of schizosaccharomyces pombe. *Yeast*. 18: 433–44.

[6] Robinow CF. (1977) The number of chromosomes in *S. pombe*: light microscopy of stained preparations. *Genetics*. 87: 491–497.

[7] Wood V, Gwilliam R, Rajandream MA, et al. (2002) The genome sequence of Schizosaccharomyces pombe. *Nature*. 415(6874):871–80.

[8] McDowall MD, Harris MA, Lock A, et al. (2015) PomBase 2015: updates to the fission yeast database. *Nucleic Acids Res*. 43(Database issue):D656–661.

[9] http://www.ebi.ac.uk/2can/genomes/eukaryotes/Schizosaccharomyces_pombe.html

[10] Osterwalder A. (1924) *Schizosaccharomyces liquefaciens* n. sp., eine gegen freie schweflige Saure widerstandsfahige Garhefe. *Mitteilungen aus dem Gebiete der Lebensmitteluntersuchung und Hygiene.* 15: 5–28.

[11] Leupold U. (1950) Die Verebung von Homothallie und Heterothallie bei Schizosaccharomyces pombe. *C.R. Lab. Carlsberg*. 24: 381–475.

[12] Gutz H, Doe FJ. (1973) Two different h- mating types in Schizosaccharomyces pombe. *Genetics*. 74: 563–569.

[13] James A. Barnett. (2007) A history of research on yeasts 10: foundations of yeast genetics 1. *Yeast*. 24: 799–845.

[14] Kohli J, Hottinger H, Munz P, et al. (1977) Genetic mapping in Schizosaccharomyces pombe by mitotic and meiotic analysis and induced haploidization. *Genetics*. 87: 471–489.

[15] Kelly M, Burke J, Smith M, et al. (1988) Four mating-type genes control sexual differentiation in the fission yeast. *EMBO J*. 7(5):1537–47.

[16] Furuya K, Niki H. (2009) Isolation of heterothallic haploid and auxotrophic mutants of Schizosaccharomyces japonicus. *Yeast*. 26(4): 221–33.

[17] Lee MG, Nurse P. (1987) Complementation used to clone a human homologue of the fission yeast cell cycle control gene cdc2+. *Nature*. 327: 31–35.

[18] Wixon J. (2002) Featured Organism: Schizosaccharomyces pombe, the fission yeast. *Comp Funct Genom*. 3: 194–204.

[19] Nurse P, Thuriaux P, Nasmyth K. (1976) Genetic control of the cell division cycle in the fission yeast Schizosaccharomyces pombe. *Molecular and General Genetics*. 146: 167–178.

[20] Wood E, Nurse P. (2013) Pom1 and cell size homeostasis in fission yeast. *Cell Cycle*. 12(19): 3228–36.

[21] Egel R. (1973. Genes involved in mating type expression of fission yeast. *Mol Gen Genet*. 122: 339–343.

[22] Tanaka K, Davey J, Imai Y, et al. (1993) Schizosaccharomyces pombe map3+ encodes the putative M-factor receptor. *Mol Cell Biol*. 13: 80–88.

[23] Mitchison JM, Nurse P. (1985) Growth in cell length in the fission yeast schizosaccharomyces pombe. *J Cell Sci*, 75: 357–376.

[24] Martin SG, Chang F. (2005) New end take off: regulatin cell polarity during the fission yeast cell cycle. *Cell Cycle*. 4: 1046–1049.

[25] Wixon J. (2002). Featured Organism: Schizosaccharomyces pombe, the fission yeast. *Comp Funct Genom*. 3: 194–204.

[26] Zhao Y, Lieberman HB. (1995) Schizosaccharomyces pombe: a model for molecular studies of eukaryotic genes. *DNA Cell Biol*. 14(5): 359–71.

[27] Idiris A, Bi K, Tohda H, et al. (2006) Construction of a protease-deficient strain set for the fission yeast Schizosaccharomyces pombe, useful for effective production of protease-sensitive heterologous proteins. *Yeast*. 23: 83–99.

[28] Barnett JA, Robinow CF. (2002) A history of research on yeasts 4: cytology part II, 1950–19901. *Yeast*. 19: 745–772.

[29] Fisher D, Nurse P. (1995) Cyclins of the fission yeast Schizosaccharomyces pombe. *Semin Cell Biol*. 6(2): 73–8.

[30] Nurse P. (2000) A long twentieth century of the cell cycle and beyond. *Cell*. 100:71–78.

[31] Mitchison JM. (1957) The growth of single cells. I. Schizosaccharomyces pombe. *Exp. Cell Res*. 13: 244–262.

[32] Hayles J, Wood V, Jeffery L, et al. (2013) A genome-wide resource of cell cycle and cell shape genes of fission yeast. *Open Biol*. 3(5): 130053.

[33] Mata J, Lyne R, Burns G, et al. (2002) The transcriptional program of meiosis and sporulation in fission yeast. *Nat Genet*. 32(1):143–7.

[34] Goya A, Takaine M, Simanis V, et al. (2011) Dividing the spoils of growth and the cell cycle: The fission yeast as a model for the study of cytokinesis. *Cytoskeleton* (Hoboken). 68(2): 69–88.

[35] Wixon J. (2002) Featured Organism: Schizosaccharomyces pombe, the fission yeast. *Comp Funct Genom*. 3: 194–204.

[36] Roux AE, Chartrand P, Ferbeyre G, et al. (2010) Fission yeast and other yeasts as emergent models to unravel cellular aging in eukaryotes. *J Gerontol A Biol Sci Med Sci*. 65(1): 1–8.

[37] Kaeberlein M, Burtner CR, Kennedy BK. (2007) Recent developments in yeast aging. *PLoS Genet*. 3(5): e84.

[38] Fabrizio P1, Longo VD. (2008) Chronological aging-induced apoptosis in yeast. *Biochim Biophys Acta*. 1783(7): 1280–1285.

[39] Roux AE, Arseneault G, Chartrand P, et al. (2010) A screen for genes involved in respiration control and longevity in Schizosaccharomyces pombe. *Ann NY Acad Sci*. 1197:19–27.

[40] Narasimhan SD, Yen K, Tissenbaum HA. (2009) Converging pathways in lifespan regulation. *Curr Biol*. 19: R657–666.

[41] Parrella E, Longo VD. (2010) Insulin/IGF-I and related signaling pathways regulate aging in nondividing cells: from yeast to the mammalian brain. *Scientific World Journal*. 10: 161–177.

[42] Bai X, Jiang Y. (2010) Key factors in mTOR regulation. *Cell Mol Life Sci*. 67: 239–253.

[43] Bhaskar PT, Hay N. (2007) The two TORCs and Akt. *Dev Cell*. 12: 487–502.

[44] Harrison DE, Strong R, Sharp ZD, et al. (2009) Rapamycin fed late in life extends life-span in genetically heterogeneous mice. *Nature*. 460: 392–395.

[45] Stanfel MN, Shamieh LS, Kaeberlein M, et al. (2009) The TOR pathway comes of age. *Biochim Biophys Acta*. 1790: 1067–1074.

[46] Kaeberlein M, Powers RW, 3rd, Steffen KK, et al. (2005) Regulation of yeast replicative life span by TOR and Sch9 in response to nutrients. Science 310: 1193–1196.

[47] Mutoh N, Kitajima S. (2007) Accelerated chronological aging of a mutant fission yeast deficient in both glutathione and superoxide dismutase having Cu and Zn as cofactors and its enhancement by sir2 deficiency. *Biosci Biotechnol Biochem*. 71: 2841–2844.

[48] Zuin A, Gabrielli N, Calvo IA, Garcia-Santamarina S, Hoe KL, et al. (2008) Mitochondrial dysfunction increases oxidative stress and decreases chronological life span in fission yeast. *PLoS ONE*. 3: e2842.

[49] Fabrizio P, Pozza F, Pletcher SD, et al. (2001) Regulation of longevity and stress resistance by Sch9 in yeast. *Science*. 292: 288–290

[50] Ravikumar B, Sarkar S, Davies JE, et al. (2010) Regulation of mammalian autophagy in physiology and pathophysiology. *Physiol Rev*. 90: 1383–1435.

[51] He C and Klionsky D. (2009) Regulation Mechanisms and Signaling Pathways of Autophagy. Annual Review of *Genetics*. 43: 67–93.

[52] Cheng Y, Ren X, Hait WN, Yang Jin-Ming. (2013)Therapeutic targeting of autophagy in disease: Biology and pharmacology. *Pharmacological Reviews*. 65: 1162–1197.

[53] Murrow L and Debnath J. (2013) Autophagy as a stress-response and quality-control mechanism: Implications for cell injury and human disease. *The Annual Review of Pathology: Mechanisms of Disease*. 8: 105–37.

[54] Kundu M and Thompson CB. (2008) Autophagy: Basic principles and relevance to disease. *Annual Review of Pathology: Mechanisms of Disease*. 3:427–55.

[55] Maycotte P and Thorburn A. (2014) Targeting autophagy in breast cancer. *World Journal of Clinical Oncology*, 5(3): 224–240.

[56] Janji B, Viry E, Baginska J, et al. (2013) Role of autophagy in cancer and tumor progression. In: Bailly Y (Ed.), *Autophagy — A Double-Edged Sword — Cell Survival or Death?*: InTech, Available from: http://www.intechopen.com/books/autophagy-a-double-edged-sword-cell-survival-or-death-/role-of-autophagy-in-cancer-and-tumor-progression (accessed 2 January 2015)

[57] de Reuck AVS, Cameron MP. *Ciba Foundation Symposium on Lysosomes*. London: J.A. Churchill Ltd, 1963.

[58] Clark SL. (1957) Cellular differentiation in the kidneys of newborn mice studied with the electron microscope. *Journal of Biophysical and Biochemical Cytology*. 3(3): 349–376.

[59] Klionsky DJ, Abdalla AF, Abeliovich H, et al. (2012) Guidelines for the use and interpretation of assays for monitoring autophagy. *Autophagy*. 8(4): 445–544.

[60] Nakatogawa H, Suzuki K, Kamada Y, Ohsumi Y. (2009) Dynamics and diversity in autophagy mechanisms: lessons from yeast. *Nature Reviews Molecular Cell Biology*. 10(7):458–67.

[61] Reggiori F and Klionsky DJ. (2002) Autophagy in the eukaryotic cell. *Eukaryotic Cell*. 1(1): 11–21.

[62] Cuervo AM and Wong E. (2014) Chaperone-mediated autophagy: roles in disease and aging. *Cell Research*. 24:92–104.

[63] Sahu R, Kaushik S, Clement CC, et al. (2011) Microautophagy of cytosolic proteins by late endosomes. *Developmental Cell*. 20(1): 131–139.

[64] Juenemann K and Reits EA. (2012) Alternative macroautophagic pathways. *International Journal of Cell Biology*. 2012:189794.

[65] Fujiwara Y, Kikuchi H, Aizawa S, et al. (2013) Discovery of a novel type of autophagy targeting RNA. *Autophagy*. 9:3: 403–409.

[66] Hemmings BA and Restuccia DF. (2012) PI3K-PKB/Akt Pathway. Cold Spring Harbor Perspectives in *Biology*. 4(9):a011189.

[67] Efeyan A, Zoncu R, Sabatini DM. (2012) Amino acids and mTORC1: from lysosomes to disease. *Trends Mol Med*. 18(9):524–33.

[68] Jung CH, Ro Seung-Hyun, Cao J, et al. (2010) mTOR regulation of autophagy. *FEBS Lett*. 584(7):1287–95

[69] Itakura E, Kishi C, Inoue K, et al. (2008) Beclin 1 Forms Two Distinct Phosphatidylinositol 3-Kinase Complexes with Mammalian Atg14 and UVRAG. *Molecular Biology of the Cell*. 19: 5360–5372.

[70] Liang C, Lee Jong-soo, Inn Kyung-Soo, et al. (2008) Beclin1-binding UVRAG targets the class C Vps complex to coordinate autophagosome maturation and endocytic trafficking. *Nature Cell Biology*. 10(7): 776–787.

[71] Klionsky DJ, Baehrecke EH, Brumell JH, et al. (2011) A comprehensive glossary of autophagy-related molecules and processes (2nd edition). *Autophagy*. 7: 1273–1294.

[72] Sun L-L, Li M, Suo F, et al. (2013) Global Analysis of Fission Yeast Mating Genes Reveals New Autophagy Factors. *PLoS Genet*. 9(8):e1003715.

[73] Mukaiyama H, Kajiwara S, Hosomi A, et al. (2009) Autophagy-deficient *Schizosaccharomyces pombe* mutants undergo partial sporulation during nitrogen starvation. *Microbiology*. 155: 3816–3826.

[74] Kohda TA, Tanaka K, Konomi M, et al. (2007) Fission yeast autophagy induced by nitrogen starvation generates a nitrogen source that drives adaptation processes. *Genes to Cells*. 12: 155–170.

[75] Flanagan MD, Whitehall SK, Morgan BA. (2013) An Atg10-like E2 enzyme is essential for cell cycle progression but not autophagy in *Schizosaccharomyces pombe*. *Cell Cycle*. 12(2): 271–277.

[76] Obara K, Sekito T, Niimi K, Ohsumi Y. (2008) The Atg18-Atg2 Complex is recruited to autophagic membranes via Phosphatidylinositol 3-Phosphate and exerts an essential function. *Journal of Biological Chemistry*. 283(35): 23972–23980.

[77] Geng J, Klionsky DJ. (2008) The Atg8 and Atg12 ubiquitin-like conjugation systems in macroautophagy. *EMBO reports*. 9:859–864.

[78] Hanada T, Noda NN, Satomi Y, et al. (2007) The Atg12-Atg5 Conjugate has a novel E3-like activity for protein lipidation in autophagy. *Journal of Biological Chemistry*. 282(52): 37298–37302.

[79] Velikkakath AKG, Nishimura T, et al. (2012) Mammalian Atg2 proteins are essential for autophagosome formation and important for regulation of size and distribution of lipid droplets. *Mol Biol Cell*. 23(5):896–909.

[80] Takahara T, Maeda T. (2012) TORC1 of fission yeast is rapamycin-sensitive. *Genes to Cells*. 17: 698–708.

[81] Hayashi T, Hatanaka M, Nagao K, et al. (2007) Rapamycin sensitivity of the *Schizosaccharomyces pombe* tor2 mutant and organization of two highly phosphorylated TOR complexes by specific and common subunits. *Genes to Cells*. 12: 1357–1370.

[82] Hosokawa N, Sasaki T, Iemura Shun-ichiro, et al. (2009) Atg101, a novel mammalian autophagy protein interacting with Atg13. *Autophagy*. 5(7): 973–979.

[83] Lockshin RA, Williams CM. (1964) Programmed Cell Death–II. Endocrine potentiation of the breakdown of the intersegmental muscles of silkmoths. *Journal of Insect Physiology*. 10: 643–9.

[84] Lockshin, RA, Zakeri Z. (2001) Programmed Cell Death and Apoptosis: Origins of the Theory. *Nature Reviews Molecular Cell Biology*. 2: 545–50.

[85] Kerr JFR, Wyllie AH, Currie AR. (1972) Apoptosis: a basic biological phenomenon with wideranging implications in tissue kinetics. *British Journal of Cancer*. 26: 239–257.

[86] Galluzzi L, Vitale I, Abrams JM, et al. (2012) Molecular definitions of cell death subroutines: recommendations of the Nomenclature Committee on Cell Death. *Cell Death and Differentiation*. 19: 107–120

[87] Brenner S. (1974) The genetics of *Caenorhabditis elegans*. *Genetics*. 77: 71–94.

[88] Kaletta T, Hengartner MO. (2006) Finding function in novel targets: *C. elegans* as a model organism. *Nat Rev Drug Discov*. 5(5): 387–98.

[89] Kinet MJ, Shaham S. (2014) Noncanonical cell death in the nematode *Caenorhabditis Elegans*. Methods in Enzymology. 545: 157–180.

[90] Conradt B and Xue D. Programmed cell death 2005, *WormBook,* ed. The *C. elegans* Research Community, WormBook, http://www.wormbook.org

[91] Kinchen JM. (2010) A model to die for: signaling to apoptotic cell removal in worm, fly and mouse. *Apoptosis*. 15:998–1006.

[92] Arvanitis M, De-Dong L, Lee K, et al. (2013) Apoptosis in *C.elegans*: lessons for cancer and immunity. *Front Cell Infect Microbiol*. 3:67.

[93] Hongmei Z. (2012) Extrinsic and intrinsic apoptosis signal pathway review, apoptosis and medicine, Ntuli T (Ed.), InTech 2012. doi:10.5772/50129. Available from: http://www.intechopen.com/books/apoptosis-and-medicine/extrinsic-and-intrinsic-apoptosis-signal-pathway-review

[94] Lamkanfi M, Dixit VM. (2010) Manipulation of host cell death pathways during microbial infections. *Cell Host Microbe*. 8(1):44–54.

[95] Elmore S. (2007) Apoptosis: A Review of Programmed Cell Death. *Toxicologic Pathology*. 35: 495–516.

[96] Taylor RC, Cullen SP, Martin SJ. (2008) Apoptosis: controlled demolition at the cellular level. *Molecular Cell Biology*. 9: 231–241.

[97] Chaabane W, User SD, El-Gazzah M, et al. (2013) Autophagy, apoptosis, mitoptosis and necrosis: Interdependence between those pathways and effects on cancer. *Archivum Immunologiae et Therapiae Experimentalis*. 61:43–58.

[98] Ow Yong-Ling, Green DR, Hao Z, et al. (2008) Cytochrome c: functions beyond respiration. *Molecular Cell Biology*. 9: 532–542.

[99] Yuan S, Akey CW. (2013) Apoptosome structure, assembly and procaspase activation. *Structure*. 21(4): 501–515.

[100] Berthelet J, Dubrez L. (2013) Regulation of Apoptosis by Inhibitors of Apoptosis (IAPs). *Cells*. 2: 163–187.

[101] Ink B, Rnig MZ, Baum B, et al. (1997) Human Bak induces cell death in *Schizosaccharomyces pombe* with morphological changes similar to those with apoptosis in mammalian cells. *Molecular and Cellular Biology*. 17(5): 2468–74.

[102] Jurgensmeier JM, Krajewski S, Armstrong RC, et al. (1997) Bax- and Bak-induced cell death in the fission yeast *Schizosaccharomyces pombe. Molecular Biology of the Cell*. 8: 325–339.

[103] James C, Gschmeissner S, Fraser A, et al. (1997) CED-4 induces chromatin condensation in *Schizosaccharomyces pombe* and is inhibited by direct physical association with CED-9. *Current Biology*. 7(4): 246–52.

[104] Mutoh N, Kitajima S, Ichihara S. (2011) Apoptotic cell death in the fission yeast *Schizosaccharomyces pombe* induced by valproic acid and its extreme susceptibility to pH change. *Bioscience, Biotechnology, and Biochemistry*. 75 (6): 1113–1118.

[105] Choon PL, Hongyuan Y. (2008) Programmed cell death in fission yeast Schizosaccharomyces pombe. Biochimica et Biophysica Acta. 1783: 1335–1349.

[106] Fröhlich Kai-Uwe, Fussi H, Ruckenstuhl C. (2007) Yeast apoptosis — From genes to pathways. *Seminars in Cancer Biology*. 17: 112–121.

[107] Carmona-Gutierrez D, Eisenberg T, Buttner S, et al. (2010) Apoptosis in yeast: triggers, pathways, subroutines. *Cell Death and Differentiation*. 17: 763–773.

[108] Volkmer E, Karnitz LM. (1999) Human homologs of *Schizosaccharomyces pombe* Rad1, Hus1, and Rad9 Form a DNA damage-responsive protein complex. *Journal of Biological Chemistry*. 274 (2): 567–570.

[109] Komatsu K, Hopkins KM, Lieberman HB, et al. (2000) *Schizosaccharomyces pombe* Rad9 contains a BH3-like region and interacts with the anti-apoptotic protein Bcl-2. *FEBS Letters*. 481: 122–126.

[110] Hassine SB, Arcangioli B. (2009) Tdp1 protects against oxidative DNA damage in non-dividing fission yeast. The European Molecular Biology Organization Journal. 28: 632–640.

[111] Burhans WC, Weinberger M, Marchetti MA, et al. (2003) Apoptosis-like yeast cell death in response to DNA damage and replication defects. *Mutation Research*. 532: 227–243.

[112] Farrugia G, Balzan R. Oxidative stress and programmed cell death in yeast. *Front Oncol*. 2:64.

[113] Low CP, Liew LP, Pervaiz S, et al. (2005) Apoptosis and lipoapoptosis in the fission yeast *Schizosaccharomyces pombe*. *FEMS Yeast Research*. 5: 1199–1206.

[114] Lim Hye-Won, Kim Su-Jung, Park Eun-Hee, et al. (2007) Overexpression of a metacaspase gene stimulates cell growth and stress response in *Schizosaccharomyces pombe*. *Canadian Journal of Microbiology*. 53(8): 1016–1023.

[115] Fahrenkrog B, Sauder U, Aebi U. (2003) The S. cerevisiae HtrA-like protein Nma111p is a nuclear serine protease that mediates yeast apoptosis. *Journal of Cell Science*. 117: 115–126.

[116] Ludovico P, Rodrigues F, Almeida A, et al. (2002) Cytochrome *c* release and mitochondria involvement in programmed cell death induced by acetic acid in *Saccharomyces cerevisiae*. *Molecular Biology of the Cell*. 13(8): 2598–2606.

[117] Zhao Y, Lieberman HB. (1995) Schizosaccharomyces pombe: a model for molecular studies of eukaryotic genes. *DNA Cell Biol*. 14(5): 359–371.

[118] Jahic M, Gustavsson M, et. al. (2003). Analysis and control of proteolysis of a fusion protein in Pichia pastoris fed-batch processes. *J Biotechnol*. 102(1): 45–53.

[119] Hirashima K, Iwaki T, Takegawa K, et al. (2006) A simple and effective chromosome modification method for large-scale deletion of genome sequences and identification of essential genes in fission yeast. *Nucleic Acids Res*. 34(2):e11.

[120] Idiris A, Tohda H, Bi KW, et al. (2006) Enhanced productivity of protease-sensitive heterologous proteins by disruption of multiple protease genes in the fission yeast Schizosaccharomyces pombe. *Appl Microbiol Biotechnol*. 73(2):404–420.

[121] Takegawa K, Tohda H, Sasaki M, et al. (2009) Production of heterologous proteins using the fission-yeast (Schizosaccharomyces pombe) expression system. *Biotechnol Appl Biochem*. 53(Pt 4):227–35.

[122] Herrera-Camacho I, Rosas-Murrieta N, Rojo-Dominguez A, (2007) Biochemical characterization and structural prediction of a novel cytosolic leucyl aminopeptidase of the M17 family from Schizosaccharomyces pombe. *FEBS Journal*. 274: 6228–6240.

[123] Suárez P, Villa L, Arbesú MJ, et al. (1991). The proteolytic system of the fission yeast Schizosaccharomyces pombe. FEMS Microbiol Lett. 65(2):215–20.

[124] Rawlings ND, Morton FR, Barrett AJ. (2006). MEROPS: the peptidase database. Nucleic Acids Res. 34: D270–D272.